Phrase Mining from
Massive Text and Its Applications

Synthesis Lectures on Data Mining and Knowledge Discovery

Editors
Jiawei Han, *University of Illinois at Urbana-Champaign*
Lise Getoor, *University of California, Santa Cruz*
Wei Wang, *University of California, Los Angeles*
Johannes Gehrke, *Cornell University*
Robert Grossman, *University of Chicago*

Synthesis Lectures on Data Mining and Knowledge Discovery is edited by Jiawei Han, Lise Getoor, Wei Wang, Johannes Gehrke, and Robert Grossman. The series publishes 50- to 150-page publications on topics pertaining to data mining, web mining, text mining, and knowledge discovery, including tutorials and case studies. Potential topics include: data mining algorithms, innovative data mining applications, data mining systems, mining text, web and semi-structured data, high performance and parallel/distributed data mining, data mining standards, data mining and knowledge discovery framework and process, data mining foundations, mining data streams and sensor data, mining multi-media data, mining social networks and graph data, mining spatial and temporal data, pre-processing and post-processing in data mining, robust and scalable statistical methods, security, privacy, and adversarial data mining, visual data mining, visual analytics, and data visualization.

Probabilistic Approaches to Recommendations
Nicola Barbieri, Giuseppe Manco, and Ettore Ritacco
2014

Outlier Detection for Temporal Data
Manish Gupta, Jing Gao, Charu Aggarwal, and Jiawei Han
2014

Provenance Data in Social Media
Geoffrey Barbier, Zhuo Feng, Pritam Gundecha, and Huan Liu
2013

Graph Mining: Laws, Tools, and Case Studies
D. Chakrabarti and C. Faloutsos
2012

Mining Heterogeneous Information Networks: Principles and Methodologies
Yizhou Sun and Jiawei Han
2012

Privacy in Social Networks
Elena Zheleva, Evimaria Terzi, and Lise Getoor
2012

Community Detection and Mining in Social Media
Lei Tang and Huan Liu
2010

Ensemble Methods in Data Mining: Improving Accuracy Through Combining Predictions
Giovanni Seni and John F. Elder
2010

Modeling and Data Mining in Blogosphere
Nitin Agarwal and Huan Liu
2009

Phrase Mining from Massive Text and Its Applications
Jialu Liu, Jingbo Shang, and Jiawei Han

ISBN: 978-3-031-00782-8   paperback
ISBN: 978-3-031-01910-4   ebook

DOI 10.1007/978-3-031-01910-4

A Publication in the Springer series
SYNTHESIS LECTURES ON DATA MINING AND KNOWLEDGE DISCOVERY

Lecture #13
Series Editors: Jiawei Han, *University of Illinois at Urbana-Champaign*
                Lise Getoor, *University of California, Santa Cruz*
                Wei Wang, *University of California, Los Angeles*
                Johannes Gehrke, *Cornell University*
                Robert Grossman, *University of Chicago*
Series ISSN
Print 2151-0067   Electronic 2151-0075

Phrase Mining from
Massive Text and Its Applications

Jialu Liu
Google

Jingbo Shang
University of Illinois at Urbana-Champaign

Jiawei Han
University of Illinois at Urbana-Champaign

*SYNTHESIS LECTURES ON DATA MINING AND KNOWLEDGE
DISCOVERY #13*

ABSTRACT

A lot of digital ink has been spilled on "big data" over the past few years. Most of this surge owes its origin to the various types of unstructured data in the wild, among which the proliferation of text-heavy data is particularly overwhelming, attributed to the daily use of web documents, business reviews, news, social posts, etc., by so many people worldwide. A core challenge presents itself: How can one efficiently and effectively turn massive, unstructured text into structured representation so as to further lay the foundation for many other downstream text mining applications?

In this book, we investigated one promising paradigm for representing unstructured text, that is, through automatically identifying high-quality phrases from innumerable documents. In contrast to a list of frequent n-grams without proper filtering, users are often more interested in results based on variable-length phrases with certain semantics such as scientific concepts, organizations, slogans, and so on. We propose new principles and powerful methodologies to achieve this goal, from the scenario where a user can provide meaningful guidance to a fully automated setting through distant learning. This book also introduces applications enabled by the mined phrases and points out some promising research directions.

KEYWORDS

phrase mining, phrase quality, phrasal segmentation, distant supervision, text mining, real-world applications, efficient and scalable algorithms

Contents

Acknowledgments

The authors would like to acknowledge Xiang Ren, Fangbo Tao, and Huan Gui for their contribution to Chapter 4.

The research was supported in part by the U.S. Army Research Lab. under Cooperative Agreement No. W911NF-09-2-0053 (NSCTA), National Science Foundation IIS-1320617, and IIS 16-18481, and grant 1U54GM114838 awarded by NIGMS through funds provided by the trans-NIH Big Data to Knowledge (BD2K) initiative (www.bd2k.nih.gov). The views and conclusions contained in this document are those of the author(s) and should not be interpreted as representing the official policies of the U.S. Army Research Laboratory or the U.S. Government. The U.S. Government is authorized to reproduce and distribute reprints for Government purposes notwithstanding any copyright notation hereon. The views and conclusions contained in our research publications are those of the authors and should not be interpreted as representing any funding agencies.

Jialu Liu, Jingbo Shang, and Jiawei Han
February 2017

CHAPTER 1

Introduction

1.1 MOTIVATION

The past decade has witnessed the surge of interest in data mining which is broadly construed to discover knowledge from all kinds of data, be it in academia, industry, or daily life. The information explosion brings the "big data" era to the light of the stage. This overwhelming tide of information is largely composed of *unstructured data* such as images, speeches, and videos. It is easy to distinguish them from typical *structured data* (e.g., relational data) in that the latter can be readily stored in the fielded form in databases. Among the various unstructured data, a particularly prominent category comes in the form of text. Examples include news articles, social media messages, as well as web pages and query logs.

In the literature of text mining, during the process of analyzing text, one fundamental problem is how to effectively represent text and model its topic, not only from the perspective of algorithm performance, but also for analysts to better interpret and present the results. A common approach is to use n-gram, i.e., a contiguous sequence of n unigrams, as the basic units. Figure 1.1 shows an example sequence with the corresponding 1-gram, 2-gram, 3-gram and consolidated representation. However, such representation raises concerns of exponential growth of the dictionary as well as the lack of interpretability. One can reasonably expect an intelligent method that only uses a compact subset of n-grams but generates explainable representation given a document.

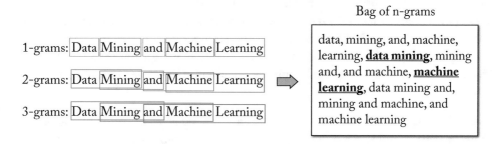

Figure 1.1: Example of n-gram representation.

Along this line of thought, in this book, we formulate such explainable n-gram subset as *quality phrases* (e.g., scientific terms such as "data mining" and "machine learning" outlined in the figure) and *phrase mining* as the corresponding knowledge discovery process.

Phrase mining has been studied in different communities. The natural language processing (NLP) community refers to it as "automatic term recognition" (i.e., extracting technical terms with the use of computers). The information retrieval (IR) community studies this topic to select main concepts in a corpus in an effort to improve search engine. Among existing works published by these two communities, linguistic processors with heuristic rules are primarily used and the most common approach is based on noun phrases. Supervised noun phrase chunking techniques are particularly proposed to leverage annotated documents to learn these rules. Other methods may utilize more sophisticated NLP features, such as dependency parser to further enhance the precision. However, emerging textual data, such as social media messages, can deviate from rigorous language rules. Using various kinds of heavily (pre-)trained linguistic processing makes these approaches difficult to be generalized.

In this regard, we believe that the community would welcome and benefit from a set of *data-driven* algorithms that work for *large-scale* datasets involving irregular textual data in a robust way, while minimizing the human labeling cost. We are also convinced by various study and experiments that our proposed methods embody enough novelty and contribution to add solid building block for various text-related tasks including document indexing, keyphrase extraction, topic modeling, knowledge base construction, and so on.

1.2   WHAT IS PHRASE MINING?

Phrase mining is a text mining technique that discovers semantically meaningful phrases from massive text. By considering the challenge of heterogeneity in the emerging textual data, the principles and methods discussed in this book will not assume particular lexical rules and are primarily compelled by data. Formally, we define the task as follows.

Problem 1.1   Phrase Mining   Given a large document corpus $\mathcal{C}$—which can be any textual word sequences with arbitrary lengths such as articles, titles, and queries—phrase mining tries to assign a value between 0 and 1 to indicate the quality of each phrase mentioned in D and discovers a set of quality phrases $K = \{K_1, \cdots, K_M\}$ with their quality scores greater than 0.5. It also seeks to provide a segmenter for locating quality phrase mentions in any unseen text snippet.

Definition 1.2   Quality Phrase.   A quality phrase is a sequence of words that appear contiguously in the corpus, and serves as a complete (non-composible) semantic unit in certain context among given documents.

There is no universally accepted definition for phrase quality. However, it is useful to quantify phrase quality based on certain criteria as outlined below:

- **Popularity**: Quality phrases should occur with sufficient frequency in the given document collection.

- **Concordance**: Concordance refers to the collocation of tokens in such a frequency that is significantly higher than what is expected due to chance. A commonly used example of a phraseological-concordance is the two phrases "strong tea" and "powerful tea." One would assume that the two phrases appear in similar frequency, yet in the English language the phrase "strong tea" is considered more proper and appears with much higher frequency. Because a concordant phrase's frequency deviates from what is expected, we consider them as belonging to a whole semantic unit.

- **Informativeness**: A phrase is informative if it is indicative of a specific topic or concept. The phrase "this paper" is popular and concordant, but is not considered to be informative in the bibliographic corpus.

- **Completeness**: Long frequent phrases and their subsequences may both satisfy the above three criteria. But apparently not all of them are qualified. A quality phrase should be interpreted as a complete semantic unit in certain contexts. The phrase "vector machine" is not considered to be complete as it mostly appears with prefix word "support."

Because single-word phrases cannot be decomposed into multiple tokens, the concordance criteria is no longer definable. As an alternative, we propose the independence criteria and will introduce it in more detail in Chapter 3.

1.3 OUTLINE OF THE BOOK

The remaining chapters of the book are outlined as follows.

- **Chapter 2: Quality Phrase Mining with User Guidance** In the literature of phrase mining, earlier work focuses on efficiently retrieving recurring word sequences and ranking them according to frequency-based statistics. However, the raw frequency from the data tends to produce misleading quality assessment, and the outcome therefore is unsatisfactory. We attempt to rectify the decisive raw frequency to help discover the true quality of a phrase by examining the context of its mentions. With *limited labeling* effort from the user, the model is able to iteratively segment the corpus into non-overlapped words and phrase sequences such that: (1) the phrase quality estimated in the previous iteration guides the segmentation and (2) segmentation results rectify raw phrase frequency and improve the process of phrase quality estimation. Such an integrated framework benefits from mutual enhancement, and achieves both high quality and high efficiency.

- **Chapter 3: Automated Quality Phrase Mining** Almost all state-of-the-art methods in NLP, IR, and text mining communities require human experts at certain levels. Such reliance on manual efforts from domain experts becomes an impediment for timely analysis of massive, emerging text corpora. Besides this issue, an ideal *automated phrase mining* method is supposed to work smoothly for multiple languages with high performance in terms of precision, recall, and efficiency. We attempt to make the phrase mining automated by utilizing external knowledge bases to remove human efforts and minimize the language dependency. Modeling single-word phrases at the same time also improves the performance, especially the recall.

Since phrase mining lays the foundation for many other downstream text mining applications, we opt to devote one chapter to discuss its applications during the latest research development.

- **Chapter 4: Phrase Mining Applications** Particularly, we would like to introduce three representative applications using phrase mining results.

- The first is a statistical inference algorithm for detecting latent quality phrases topically relevant to a single document. Previously mentioned phrase mining methods are able to locate any phrase mentions in a document, but they cannot provide the relatedness between the document and the phrase.

- The second application utilizes phrase mining results to systematically analyze large numbers of textual documents from the perspective of topic exploration. We discuss how to group phrases into clusters sharing the same topic, how to summarize commonalities and differences given multiple document collections, and how to incorporate document-associated metadata like authors and tags into the exploration process.

- The last application tries to construct semantically rich knowledge base out of unstructured text. Identifying the phrases in text that constitute entity mentions and assigning types to these spans as well as to the relations between entity mentions are the key to this process.

CHAPTER 2

Quality Phrase Mining with User Guidance

In large, dynamic collections of documents, analysts are often interested in variable-length phrases, including scientific concepts, events, organizations, products, slogans, and so on. Accurate estimation of phrase quality is critical for the extraction of quality phrases and will enable a large body of applications to transform from word granularity to phrase granularity. In this chapter, we study a segmentation-integrated framework to mine multi-word quality phrases with a small set of user-provided binary labels.

2.1 OVERVIEW

Identifying quality phrases has gained increased attention due to its value of handling increasingly massive text datasets. As the origin, the natural language processing (NLP) community has conducted extensive studies mostly known as automatic term recognition [Frantzi et al., 2000, Park et al., 2002, Zhang et al., 2008], referring to the task of extracting technical terms with the use of computers. This topic also attracts attention in the information retrieval (IR) community since appropriate indexing term selection is critical to the improvement of a search engine where the ideal indexing units should represent the main concepts in a corpus, beyond the bag-of-words.

Linguistic processors are commonly used to filter out stop words and restrict candidate terms to noun phrases. With pre-defined part-of-speech (POS) rules, one can generate noun phrases as term candidates to each POS-tagged document. Supervised noun phrase chunking techniques [Chen and Chen, 1994, Punyakanok and Roth, 2001, Xun et al., 2000] leverage annotated documents to automatically learn these rules. Other methods may utilize more sophisticated NLP features such as dependency parser to further enhance the precision [Koo et al., 2008, McDonald et al., 2005]. With candidate terms collected, the next step is to leverage certain statistical measures derived from the corpus to estimate phrase quality. Some methods further resort to reference corpus for the calibration of "termhood" [Zhang et al., 2008]. The various kinds of linguistic processing, domain-dependent language rules, and expensive human labeling make it challenging to apply the phrase mining technique to emerging big and unrestricted corpora which possibly encompass many different domains and topics such as query logs, social media messages, and textual transaction records. Therefore, researchers have sought more general data-driven approaches, primarily based on the frequent pattern mining principle [Ahonen, 1999, Simitsis et al., 2008]. Early work focuses on efficiently retrieving recurring word sequences, but many such sequences do not form meaningful phrases. More recent work filters or ranks them according to frequency-based statistics. However, the raw frequency from the data tends to produce misleading quality assessment, and the outcome is unsatisfactory, as the following example demonstrates.

Example 2.1 Raw Frequency-based Phrase Mining Consider a set of scientific publications and the raw frequency counts of two phrases "relational database system" and "support vector machine" and their subsequences in the *frequency* column of Table 2.1. The numbers are hypothetical but manifest several key observations: (i) the frequency generally decreases with the phrase length; (ii) both good and bad phrases can possess high frequency (e.g., "support vector" and "vector machine"); and (iii) the frequency of one sequence (e.g., "relational database system") and its subsequences can have a similar scale of another sequence (e.g., "support vector machine") and its counterparts.

Table 2.1: A hypothetical example of word sequence raw frequency

Sequence	Raw Frequency	Quality Phrase?	Rectified Frequency
relational database system	100	yes	70
relational database	150	yes	40
database system	160	yes	35
relational	500	N/A	20
database	1000	N/A	200
system	10000	N/A	1000
Sequence	**Raw Frequency**	**Quality Phrase?**	**Rectified Frequency**
support vector machine	100	yes	80
support vector	160	yes	50
vector maching	150	no	6
support	500	N/A	150
vector	1000	N/A	200
machine	10000	N/A	50

Obviously, a method that ranks the word sequences solely according to the frequency will output many false phrases such as "vector machine." In order to address this problem, different heuristics have been proposed based on comparison of a sequence's frequency and its sub- (or super-) sequences, assuming that a good phrase should have high enough (normalized) frequency compared with its sub-sequences and/or super-sequences [Danilevsky et al., 2014, Parameswaran et al., 2010]. However, such heuristics can hardly differentiate the quality of, e.g., "support vector" and "vector machine" because their frequencies are so close. Finally, even if the heuristics can indeed draw a line between "support vector" and "vector machine" by discriminating their frequencies (between 160 and 150), the same separation could fail for another case like "relational database" and "database system."

Using the frequency in Table 2.1, all heuristics will produce identical predictions for "relational database" and "vector machine," guaranteeing one of them to be wrong. This example suggests the intrinsic limitations of using raw frequency counts, especially in judging whether a sequence is too long (longer than a minimum semantic unit), too short (broken and not informative), or right in length. It is a critical bottleneck for all frequency-based quality assessment.

2.2   PHRASAL SEGMENTATION

In this chapter, we discuss how to address this bottleneck through rectifying the decisive raw frequency that hinders discovering the true quality of a phrase. The goal of the *rectification* is to estimate how many times each word sequence should be interpreted in whole as a phrase in its occurrence context. The following example illustrates this idea.

Example 2.2   Rectification   Consider the following occurrences of the six multi-word sequences listed in Table 2.1.

1. A ⌈relational database system⌋ for images…

2. ⌈Database system⌋ empowers everyone in your organization…

3. More formally, a ⌈support vector machine⌋ constructs a hyperplane…

4. The ⌈support vector⌋ method is a new general method of ⌈function estimation⌋…

5. A standard ⌈feature vector⌋ ⌈machine learning⌋ setup is used to describe…

6. ⌈Relevance vector machine⌋ has an identical ⌈functional form⌋ to the ⌈support vector machine⌋…

7. The basic goal for ⌈object-oriented relational database⌋ is to ⌈bridge the gap⌋ between…

The first four instances should provide positive counts to these sequences, while the last three instances should not provide positive counts to "vector machine" or "relational database" because they should not be interpreted as a whole phrase (instead, sequences like "feature vector" and "relevance vector machine" can). Suppose one can correctly count true occurrences of the sequences, and collect rectified frequency as shown in the *rectified* column of Table 2.1. The rectified frequency now clearly distinguishes "vector machine" from the other phrases, since "vector machine" rarely occurs as a whole phrase.

The success of this approach relies on reasonably accurate rectification. Simple arithmetics of the raw frequency, such as subtracting one sequence's count with its quality super sequence, are prone to error. First, which super sequences are quality phrases is a question in and of itself.

Second, it is context-dependent to decide whether a sequence should be deemed a whole phrase. For example, the fifth instance in Example 2.2 prefers "feature vector" and "machine learning" over "vector machine," even though neither "feature vector machine" nor "vector machine learning" is a quality phrase. The context information is lost when we only collect the frequency counts.

In order to recover the true frequency with best effort, we ought to examine the context of every occurrence of each word sequence and decide whether to count it as a phrase. The examination for one occurrence may involve enumeration of alternative possibilities, such as extending the sequence or breaking the sequence, and comparison among them. The test for word sequence occurrences could be expensive, losing the advantage in efficiency of the frequent pattern mining approaches.

Facing the challenge of accuracy and efficiency, we propose a segmentation approach called "phrasal segmentation," and integrate it with the phrase quality assessment in a unified framework with linear complexity (w.r.t the corpus size). First, the segmentation assigns every word occurrence to only one phrase. In the first instance of Example 2.2, "relational database system" are bundled as a single phrase. Therefore, it automatically avoids double counting "relational database" and "database system" within this instance. Similarly, the segmentation of the fifth instance contributes to the count of "feature vector" and "machine learning" instead of "feature," "vector machine," and "learning." This strategy condenses the individual tests for each word sequence and reduces the overall complexity while ensures correctness. Second, although there are an exponential number of possible partitions of the documents, we are concerned with those relevant to the phrase extraction task only. Therefore, we can integrate the segmentation with the phrase quality assessment, such that: (i) only frequent phrases with reasonable quality are taken into consideration when enumerating partitions; and (ii) the phrase quality guides the segmentation, and the segmentation rectifies the phrase quality estimation. Such an integrated framework benefits from mutual enhancement, and achieves both high quality and high efficiency.

A phrasal segmentation defines a partition of a sequence into subsequences, such that every subsequence corresponds to either a single word or a phrase. Example 2.2 shows instances of such partitions, where all phrases with high quality are marked by brackets ⌈⌋. The phrasal segmentation is distinct from word, sentence or topic segmentation tasks in natural language processing. It is also different from the syntactic or semantic parsing which relies on grammar to decompose the sentences with rich structures like parse trees. Phrasal segmentation provides the necessary granularity we need to extract quality phrases. The total count for a phrase to appear in the segmented corpus is called *rectified frequency*.

It is beneficial to acknowledge that a sequence's segmentation may not be unique, for two reasons. First, as we mentioned above, a word sequence may be regarded as a phrase or not, depending on the adoption customs. Some phrases, like "bridge the gap" in the last instance of Example 2.2, are subject to a user's requirement. Therefore, we seek for segmentation that accommodates the phrase quality, which is learned from user-provided examples. Second, a sequence could be ambiguous and have different interpretations. Nevertheless, in most cases, it does not require perfect segmentation, no matter if such a segmentation exists, to extract quality phrases. In a large document collection, the popularly adopted phrases appear many times in a variety of context. Even with a few mistakes or debatable partitions, a reasonably high quality segmentation (e.g., yielding no partition like "support ⌈vector machine⌋") would retain sufficient support (i.e., rectified frequency) for these quality phrases, albeit not for false phrases with high raw frequency.

With the above discussions, we have the following formalization.

Definition 2.3   Phrasal Segmentation.   Given a word sequence $C = w_1 w_2 \ldots w_n$ of length n, a segmentation $S = s_1 s_2 \ldots s_m$ for C is induced by a boundary index sequence $B = \{b_1, b_2, \ldots, b_{m+1}\}$ satisfying $1 = b_1 < b_2 < \cdots < b_{m+1} = n + 1$, where a segment $s_t = w_{b_t} w_{b_t+1} \ldots w_{b_t+|s_t|-1}$. Here $|s_t|$ refers to the number of words in segment s_t. Since $b_t + |s_t| = b_{t+1}$, for clearness we use $w_{[b_t, b_{t+1})}$ to denote word sequence $w_{b_t} w_{b_t+1} \cdots w_{b_t+|s_t|-1}$.

Example 2.4   Continuing our previous Example 2.2 and specifically for the first instance, the word sequence and marked segmentation are

$$C = \text{a relational database system for images}$$
$$S = /\text{ a }/\text{ relational database system }/\text{ for }/\text{ images }/$$

with a boundary index sequence $B = \{1, 2, 5, 6, 7\}$ indicating the location of segmentation symbol $/$.

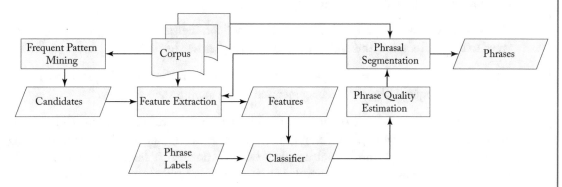

Figure 2.1: The supervised phrase mining framework.

2.3   SUPERVISED PHRASE MINING FRAMEWORK

In this chapter, in addition to the input corpus originally mentioned in Definition 1.1, users are required to provide a small set L of labeled quality phrases and $\bar{L}$ of inferior ones, which serves as the training data to guide the phrasal segmentation. The supervised framework comprises the following five steps and try to mine quality phrases following the quality criteria described in Section 1.2.

1. Generate frequent phrase candidates according to popularity criterion (Section 2.3.1).

2. Estimate phrase quality based on features design for concordance and informativeness criteria (Section 2.3.2).

3. Estimate rectified frequency via phrasal segmentation (Section 2.3.3).

4. Add segmentation-based features derived from rectified frequency into the feature set of phrase quality classifier (Section 2.3.4). Repeat steps 2 and 3.

5. Filter phrases with low rectified frequencies to satisfy the completeness criterion as post-processing step.

An complexity analysis for this framework is given at Section 2.3.5 to show that both of its computation time and required space grow linearly as the corpus size increases.

2.3.1   FREQUENT PHRASE DETECTION

The task of detecting frequent phrases can be defined as collecting aggregate counts for all phrases in a corpus that satisfy a certain minimum support threshold τ, according to the popularity criterion. In practice, one can also set a maximum phrase length ω to restrict the phrase length. Even if no explicit restriction is added, ω is typically a small constant. For efficiently mining these frequent phrases, we draw upon two properties.

1. Downward Closure property: If a phrase is not frequent, then any its super-phrase is guaranteed to be not frequent. Therefore, those longer phrases will be filtered and never expanded.

2. Prefix property: If a phrase is frequent, any of its prefix units should be frequent too. In this way, all the frequent phrases can be generated by expanding their prefixes.

The algorithm for detecting frequent phrases is given in Algorithm 1. We use $\mathcal{C}[\cdot]$ to index a word in the corpus string and $|\mathcal{C}|$ to denote the corpus size. The $\oplus$ operator is for concatecating two words or phrases. Algorithm 1 returns a key-value dictionary f. Its keys are vocabulary $\mathcal{U}$ containing all frequent phrases $\mathcal{P}$, and words $\mathcal{U} \setminus \mathcal{P}$. Its values are their raw frequency.

2.3.2   PHRASE QUALITY ESTIMATION

Estimating phrase quality from only a few training labels is challenging since a huge number of phrase candidates might be generated from the first step and they are messy. Instead of using one or two statistical measures [El-Kishky et al., 2015, Frantzi et al., 2000, Park et al., 2002], we opt to compute multiple features for each candidate in $\mathcal{P}$. A classifier is trained on these features to predict quality Q for all unlabeled phrases. For phrases not in $\mathcal{P}$, their quality is simply 0.

We divide the features into two categories according to concordance and informativeness criteria in the following two subsections. Only representative features are introduced for clearness. We then discuss about the classifier in Section 14.

Algorithm 1: Frequent Phrase Detection

---

1 **Input**: Document corpus $\mathcal{C}$, minimum support threshold τ.
2 **Output**: Raw frequency dictionary f of frequent phrases and words.
3   $f \leftarrow$ an empty dictionary
4   $index \leftarrow$ an empty dictionary
5 **for** $i \leftarrow 1$ **to** $|\mathcal{C}|$ **do**
6     $\lfloor\ index[\mathcal{C}[i]] \leftarrow index[\mathcal{C}[i]] \cup i$

7 **while** $index$ *is not empty* **do**
8     $index' \leftarrow$ an empty dictionary
9   **for** $u \in index.keys$ **do**
10   **if** $|index[u]| \geq \tau$ **then**
11            $f[u] \leftarrow |index[u]|$
12   **for** $j \in index[u]$ **do**
13                $u' \leftarrow u \oplus \mathcal{C}[j+1]$
14                $index'[u'] \leftarrow index'[u'] \cup \{j+1\}$

15     $index \leftarrow index'$

16 **return** f

Concordance Features

This set of features is designed to measure concordance among sub-units of a phrase. To make phrases with different lengths comparable, we partition each phrase candidate into two disjoint parts in all possible ways and derive effective features measuring their concordance.

Suppose for each word or phrase $u \in \mathcal{U}$, we have its raw frequency $f[u]$. Its probability $p(u)$ is defined as:

$$p(u) = \frac{f[u]}{\sum_{u' \in \mathcal{U}} f[u']}.$$

Given a phrase $v \in \mathcal{P}$, we split it into two most-likely sub-units $\langle u_l, u_r \rangle$ such that *pointwise mutual information* is minimized. Pointwise mutual information quantifies the discrepancy between the probability of their true collocation and the presumed collocation under independence assumption. Mathematically,

$$\langle u_l, u_r \rangle = \arg \min_{u_l \oplus u_r = v} \log \frac{p(v)}{p(u_l)p(u_r)}.$$

With $\langle u_l, u_r \rangle$, we directly use the pointwise mutual information as one of the concordance features.

$$PMI(u_l, u_r) = \log \frac{p(v)}{p(u_l)p(u_r)}.$$

Another feature is also from information theory, called pointwise Kullback-Leibler divergence:

$$PKL(v \| \langle u_l, u_r \rangle) = p(v) \log \frac{p(v)}{p(u_l)p(u_r)}.$$

The additional $p(v)$ is multiplied with pointwise mutual information, leading to less bias toward rare-occurred phrases.

Both features are supposed to be positively correlated with concordance.

Informativeness Features

Some candidates are unlikely to be informative because they are functional or stop words. We incorporate the following stop word-based features into the classification process.

- Whether stop words are located at the beginning or the end of the phrase candidate,

which requires a dictionary of stop words. Phrases that begin or end with stop words, such as "I am," are often functional rather than informative.

A more generic feature is to measure the informativeness based on corpus statistics:

- Average inverse document frequency (IDF) computed over words,

where IDF for a word w is computed as

$$IDF(w) = \log \frac{|\mathcal{C}|}{|\{d \in [D] : w \in C_d\}|}.$$

It is a traditional information retrieval measure of how much information a word provides in order to retrieve a small subset of documents from a corpus. In general, quality phrases are expected to have not too small average IDF.

In addition to word-based features, punctuation is frequently used in text to aid interpretations of specific concept or idea. This information is helpful for our task. Specifically, we adopt the following feature.

- Punctuation: probabilities of a phrase in quotes, brackets, or capitalized;

higher probability usually indicates more likely a phrase is informative.

It's noteworthy that in order to extract features efficiently we have designed an adapted Aho-Corasick Automaton to rapidly locate occurrences of phrases in the corpus.

The Aho-Corasick Automaton is similar to the data structure Trie, which could utilize the common prefix to save the memory usage and also make the process more efficient. It also computes the field "failed" referring to the node which could continue the matching process. In this book, we adopt standard Aho-Corasick Automaton definition and construction process. Algorithm 2 introduced a "while" loop to fix the issues brought by prefix (i.e., there might be some phrase candidates which are the prefix of the others), which is slightly different from the

---

Algorithm 2: Locating using Aho-Corasick Automaton

---

1 **Input**: The Aho-Corasick Automaton root, the corpus string C.
2 **Output**: All occurrences of phrases in the automaton $\mathcal{O}$.
3 $\mathcal{O} \leftarrow \emptyset$
4 $u \leftarrow$ root
5 **for** $i \leftarrow 1$ **to** $|C|$ **do**
6   **while** $u \neq root$ and $C[i]$ not in $u.next$ **do**
7          $u \leftarrow u.$failed
8   **if** $C[i]$ in $u.next$ **then**
9          $u \leftarrow u.$next$[C[i]]$
10      $p \leftarrow u$
11   **while** $p \neq root$ and $p.isEnd$ **do**
12          $\mathcal{O} \leftarrow \mathcal{O} \cup [i - p.depth + 1, i]$
13          $p \leftarrow p.$failed
14 **return** $\mathcal{O}$

---

traditional matching process and could help us find all occurrences of the phrase candidates in the corpus in a linear time.

An alternative way is to adopt the hash table. However, one should carefully choose the hash function for hash tale and the theoretical time complexity of hash table is not exactly linear. For comparison, we implemented a hash table approach using the unordered map in C++, while the Aho-Corasick Automaton was coded in C++ too. The results can be found in Table 2.2 We can see that Aho-Corasick Automaton is slightly better because of its exact linear complexity and less memory overhead.

Table 2.2: Runtime of locating phrase candidates

	Academia	Yelp
Aho-Corasick Automaton	154.25s	198.39s
Unordered Map (Hash Table)	192.71s	366.67s

Classifier

The framework can work with arbitrary classifiers that can be effectively trained with small labeled data and output a probabilistic score between 0 and 1. For instance, we can adopt random forest [Breiman, 2001] which is efficient to train with a small number of labels. The ratio of positive predictions among all decision trees can be interpreted as a phrase's quality estimation. In experiments we will see that 200–300 labels are enough to train a satisfactory classifier.

Just as we have mentioned, both quality phrases and inferior ones are required as labels for training. To further reduce the labeling effort, the next chapter introduces distant learning to automatically retrieve both positive and negative labels.

2.3.3   RECTIFICATION THROUGH PHRASAL SEGMENTATION

The discussion in Example 2.1 points out the limitations of using only raw frequency counts. Instead, we ought to examine the context of every word sequence's occurrence and decide whether to count it as a phrase, as introduced in Example 2.2. The segmentation directly addresses the completeness criterion, and indirectly helps with the concordance criterion via rectified frequency. Here we propose an efficient phrasal segmentation method to compute rectified frequency of each phrase. We will see that combined with aforementioned phrase quality estimation, some phrases with high raw frequency get removed as their rectified frequencies approach zero. Furthermore, rectified phrase frequencies can be fed back to generate additional features and improve the phrase quality estimation. This will be discussed in the next subsection.

The segmentation is realized through a statistical model. Given a word sequence C, and a segmentation $S = s_1 \ldots s_m$ induced by boundary index sequence $B = \{b_1, \ldots, b_{m+1}\}$, where $s_t = w_{[b_t, b_{t+1})}$, the joint probability is factorized as:

$$p(S, C) = \prod_{t=1}^{m} p\left(b_{t+1}, \lceil w_{[b_t, b_{t+1})} \rfloor \big| b_t \right), \tag{2.1}$$

where $p(b_{t+1}, \lceil w_{[b_t, b_{t+1})} \rfloor | b_t)$ is the probability of observing a word sequence $w_{[b_t, b_{t+1})}$ as the t-th quality segment. As segments of a word sequence usually have weak dependence on each other, we assume they are generated one by one for the sake of both efficiency and simplicity.

We now describe the generative model for each segment. Given the start index b_t of a segment s_t, we first generate the end index b_{t+1}, according to a prior distribution $p(|s_t| = b_{t+1} - b_t)$ over phrase lengths. Then we generate the word sequence $w_{[b_t, b_{t+1})}$ according to a multinomial distribution over all segments of length $(b_{t+1} - b_t)$. Finally, we generate an indicator whether $w_{[b_t, b_{t+1})}$ forms a quality segment according to its quality $p(\lceil w_{[b_t, b_{t+1})} \rfloor | w_{[b_t, b_{t+1})}) = Q(w_{[b_t, b_{t+1})})$. We formulate its probabilistic factorization as follows:

$$p\left(b_{t+1}, \lceil w_{[b_t, b_{t+1})} \rfloor \| b_t\right) = p\left(b_{t+1} | b_t\right) p\left(\lceil w_{[b_t, b_{t+1})} \rfloor \| b_t, b_{t+1}\right)$$

$$= p\left(b_{t+1} - b_t\right) p\left(w_{[b_t, b_{t+1})} \big\| |s_t| = b_{t+1} - b_t\right) Q\left(w_{[b_t, b_{t+1})}\right).$$

The length prior $p(|s_t| = b_{t+1} - b_t)$ is explicitly modeled to counter the bias to longer segments as they result in fewer segments. The particular form of $p(|s_t|)$ we pick is:

$$p(|s_t|) \propto \alpha^{1 - |s_t|}. \tag{2.2}$$

Here $\alpha \in R^+$ is a factor called *segment length penalty*. If $\alpha < 1$, phrases with longer length have larger value of $p(|s_t|)$. If $\alpha > 1$, the mass of $p(|s_t|)$ moves toward shorter phrases. Smaller α favors longer phrases and results in fewer segments. Tuning its value turns out to be a trade-off between precision and recall for recognizing quality phrases. At the end of this subsection we will discuss how to estimate its value by reusing labels in Section 2.3.2. It is worth mentioning that such segment length penalty is also discussed by Li et al. [2011]. Our formulation differs from theirs by posing a weaker penalty on long phrases.

We denote $p(w_{[b_t,b_{t+1})} \big| |s_t|)$ with $\theta_{w_{[b_t,b_{t+1})}}$ for convenience. For a given corpus $\mathcal{C}$ with D documents, we need to estimate $\theta_u = p(u \big| |u|)$ for each frequent word and phrase $u \in \mathcal{U}$, and infer segmentation S. Since $P(\mathcal{C})$ does not depend on segmentation S, one can maximize $\log P(S, \mathcal{C})$ instead. We employ the maximum *a posteriori* principle and maximize the joint probability of the corpus:

$$\sum_{d=1}^{D} \log p(S_d, C_d) = \sum_{d=1}^{D} \sum_{t=1}^{m_d} \log p\left(b_{t+1}^{(d)}, \lceil w_{[b_t,b_{t+1})}^{(d)} \rfloor \big| b_t^{(d)}\right). \tag{2.3}$$

To find the best segmentation to maximize Eq. (2.3), one can use efficient dynamic programming (DP) if θ is known. The algorithm is shown in Algorithm 3.

To learn θ, we employ an optimization strategy called Viterbi Training (VT) or Hard-EM in the literature [Allahverdyan and Galstyan, 2011]. Generally speaking, VT is an efficient and iterative way of parameter learning for probabilistic models with hidden variables. In our case, given corpus $\mathcal{C}$, it searches for a segmentation that maximizes $p(S, \mathcal{C}|Q, \theta, \alpha)$ followed by coordinate ascent on parameters θ. Such a procedure is iterated until a stationary point has been reached. The corresponding algorithm is given in Algorithm 4.

The hard E-step is performed by DP with θ fixed, and the M-step is based on the segmentation obtained from DP. Once the segmentation S is fixed, the closed-form solution of θ_u can be derived as:

$$\theta_u = \frac{\sum_{d=1}^{D} \sum_{t=1}^{m_d} \mathbf{1}_{s_t^{(d)}=u}}{\sum_{d=1}^{D} \sum_{t=1}^{m_d} \mathbf{1}_{|s_t^{(d)}|=|u|}}, \tag{2.4}$$

where $\mathbf{1}$ denotes the identity indicator. We can see that θ_u is the rectified frequency of u normalized by the total frequencies of the segments with length $|u|$. For this reason, we name θ *normalized rectified frequency*.

Note that Soft-EM (i.e., Bawm-Welch algorithm [Bishop, 2006]) can also be applied to find a maximum likelihood estimator of θ. Nevertheless, VT is more suitable in our case because:

1. VT uses DP for the segmentation step, which is significantly faster than Bawm-Welch using forward-backward algorithm for the E-step; and

2. majority of the phrases get removed as their θ approaches 0 during iterations, which further speeds up our algorithm.

Algorithm 3: Dynamic Programming (DP)

---

1 **Input**: Word sequence $C = w_1 w_2 \ldots w_n$, phrase quality Q, normalized frequency θ,
   segment length penalty α.
2 **Output**: Optimal segmentation S.
3 $h_0 \leftarrow 1, h_i \leftarrow 0 \ (0 < i \leq n)$
4 denote ω as the maximum phrase length
   // Denote ω as the maximum phrase length.
5 **for** $i = 1$ **to** n **do**
6   **for** $\delta = 1$ **to** ω **do**
7   **if** $h_i \cdot p(b_{t+1} = b_t + \delta, \lceil w_{[i+1,i+\delta+1)} \rfloor \big| b_t) > h_{i+\delta}$ **then**
8             $h_{i+\delta} \leftarrow h_i \cdot p(b_{t+1} = b_t + \delta, \lceil w_{[i+1,i+\delta+1)} \rfloor \big| b_t)$
9             $g_{i+\delta} \leftarrow i$

10 $i \leftarrow n$
11 $m \leftarrow 0$
12 **while** $i > 0$ **do**
13     $m \leftarrow m + 1$
14     $s_m \leftarrow w_{g_i+1} w_{g_i+2} \ldots w_i$
15     $i \leftarrow g_i$
16 **return** $S \leftarrow s_m s_{m-1} \ldots s_1$

It has also been reported in Allahverdyan and Galstyan [2011] that VT converges faster and results in sparser and simpler models for Hidden Markov Model-like tasks. Meanwhile, VT is capable of correctly recovering most of the parameters.

Previously in Equation (2.2), we defined the formula of segment length penalty. There is a hyper-parameter α that needs to be determined outside the VT iterations. An overestimate α will segment quality phrases into shorter parts, while an underestimate of α tends to keep low-quality phrases. Thus, an appropriate α reflects the user's trade-off between precision and recall. To judge what α value is reasonable, we propose to reuse the labeled phrases used in the phrase quality estimation. Specifically, we try to search for the maximum value of α such that VT does not segment positive phrases. A parameter r_0 named *non-segmented ratio* controls the trade-off mentioned above. It is the expected ratio of phrases in L not partitioned by dynamic programming. The detailed searching process is described in Algorithm 5 where we initially set upper and lower bounds of α and then perform a binary search. In Algorithm 5, $|S|$ denotes the number of segments in S and $|L|$ refers to the number of positive labels.

Algorithm 4: Viterbi Training (VT)

---

1 **Input**: Corpus $\mathcal{C}$, phrase quality Q, length penalty α.

2 **Output**: θ.

3 initialize θ with normalized raw frequencies in the corpus

4 **while** *not converge* **do**

5     $\theta'_u \leftarrow 0, \forall\ u \in \mathcal{U}$

6   **for** $d = 1$ **to** D **do**

7         $S_d \leftarrow DP(C_d, Q, \theta, \alpha)$ via Algorithm 3

8   assume $S_d = s_1^{(d)} \cdots s_m^{(d)}$

9   **for** $t = 1$ **to** m **do**

10             $u \leftarrow w_{[b_t, b_{t+1})}^{(d)}$

11             $\theta'_u \leftarrow \theta'_u + 1$

12   normalize θ' w.r.t. different length as in Eq. (2.4)

13     $\theta \leftarrow \theta'$

14 **return** θ

2.3.4 FEEDBACK AS SEGMENTATION FEATURES

Rectified frequencies can help refine the features and improve the quality assessment. The motivation behind this feedback idea is explained with the examples shown in Table 2.3. "Quality before feedback" listed in the table is computed based on phrase quality estimation in the previous step. For example, the quality of "np hard in the strong" is significantly overestimated according to the raw frequency. Once we correctly segment the documents, its frequency will be largely reduced, and we can use it to guide the quality estimator. For another example, the quality of phrases like "data stream management system" were originally underestimated due to its relatively lower frequency and smaller concordance feature values. Suppose after the segmentation, this phrase is not broken into smaller units in most cases. Then we can feed that information back to the quality estimator and boost the score.

    Based on this intuition, we design two new features named *segmentation features* and plug them into the feature set introduced in Section 2.3.2. Given a phrase $v \in \mathcal{P}$, these two segmentation features are defined as:

$$\log \frac{p(S, v)|_{|S|=1}}{\max_{|S|>1} p(S, v)}$$

$$p(S, v)|_{|S|=1} \log \frac{p(S, v)|_{|S|=1}}{\max_{|S|>1} p(S, v)},$$

where $p(S, v)$ is computed by Equation (2.1). Instead of splitting a phrase into two parts like the concordance features, we now find the best segmentation with dynamic programming in-

---

Algorithm 5: Penalty Learning

1 **Input:** Corpus $\mathcal{C}$, labeled quality phrases L, phrase quality Q, non-segmented ratio r_0.
2 **Output:** Desired segment length penalty α.
3 $up \leftarrow 200, low \leftarrow 0$
4 **while** *not converge* **do**
5      $\alpha \leftarrow (up + low)/2$
6      $\theta \leftarrow VT(\mathcal{C}, Q, \alpha)$ via Algorithm 4
7      $r \leftarrow r_0 \times |L|$
8   **for** $i = 1$ *to* $|L|$ **do**
9          $S \leftarrow DP(L_i, Q, \theta, \alpha)$ via Algorithm 3
10   **if** $|S| = 1$ **then**
11             $r \leftarrow r - 1$
12   **if** $r \geq 0$ **then**
13         $up \leftarrow \alpha$
14   **else**
15         $low \leftarrow \alpha$
16 **return** $(up + low)/2$

Table 2.3: Effects of segmentation feedback on phrase quality estimation

Phrase	Quality Before	Quality After	Problem Fixed by Feedback
np hard in the strong sense	0.78	0.93	slight underestimate
np hard in the strong	0.70	0.23	overestimate
false pos. and false neg.	0.90	0.97	N/A
pos. and false neg.	0.87	0.29	overestimate
data base management system	0.60	0.82	underestimate
data stream management system	0.26	0.64	underestimate

troduced in the phrasal segmentation, which better models the concordance criterion. In addition, normalized rectified frequencies are used to compute these new features. This addresses the context-dependent completeness criterion. As a result, misclassified phrase candidates in the above example can get mostly corrected after retraining the classifier, as shown in Table 2.3.

A better phrase quality estimator can guide a better segmentation as well. In this way, the loop between the quality estimation and phrasal segmentation is closed and such an integrated framework is expected to leverage mutual enhancement and address all the four phrase quality criteria organically.

Note that we do not need to run quality estimation and phrasal segmentation for many iterations. In our experiments, the benefits brought by rectified frequency can penetrate after the first iteration, leaving performance curves over the next several iterations similar. It will be shown in the experiments.

2.3.5   COMPLEXITY ANALYSIS

Frequent Phrases Detection: Since the operation of Hash table is $\mathcal{O}(1)$, both the time and space complexities are $\mathcal{O}(\omega|\mathcal{C}|)$. ω is a small constant indicating the maximum phrase length, so this step is linear to the size of corpus $|\mathcal{C}|$.

Feature Extraction: When extracting features, the most challenging problem is how to efficiently locate these phrase candidates in the original corpus, because the original texts are crucial for finding the punctuation and capitalization information. Instead of using some dictionaries to store all the occurrences, we take the advantage of the Aho-Corasick Automaton algorithm and tailor it to find all the occurrences of phrase candidates. The time complexity is $\mathcal{O}(|\mathcal{C}| + |\mathcal{P}|)$ and space complexity $\mathcal{O}(|\mathcal{P}|)$, where $|\mathcal{P}|$ refers to the total number of frequent phrase candidates. As the length of each candidate is limited by a constant ω, $\mathcal{O}(|\mathcal{P}|) = \mathcal{O}(|\mathcal{C}|)$, so the complexity is $\mathcal{O}(|\mathcal{C}|)$ in both time and space.

Phrase Quality Estimation: As we only labeled a very small set of phrase candidates, as long as the number and depth of decision trees in the random forest are some constant, the training time for the classifier is very small compared to other parts. For the prediction stage, it is proportional to the size of phrase candidates and the dimensions of features. Therefore, it could be $\mathcal{O}(|\mathcal{C}|)$ in both time and space, although the actual magnitude might be smaller.

Viterbi Training: It is easy to observe that Algorithm 3 is $\mathcal{O}(n\omega)$, which is linear to the number of words. ω is treated as a constant, and thus the VT process is also $\mathcal{O}(|\mathcal{C}|)$ considering Algorithm 4 ususally finishes in a few iterations.

Penalty Learning: Suppose we only require a constant ϵ to check the convergence of the binary search. Then after $\log_2 \frac{200}{\epsilon}$ rounds, the algorithm converges. So the number of loops could be treated as a constant. Because VT takes $\mathcal{O}(|\mathcal{C}|)$ time, penalty learning also takes $\mathcal{O}(|\mathcal{C}|)$ time.

Summary.   Because the time and space complexities of all components in our framework are $\mathcal{O}(|\mathcal{C}|)$, our proposed framework has a linear time and space complexities and is thus very efficient. Furthermore, the most time consuming parts, including penalty learning and VT, could be easily parallelized because of the nature of independence between documents and sentences.

2.4   EXPERIMENTAL STUDY

In this section, experiments demonstrate the effectiveness and efficiency of the proposed methods in mining quality phrases and generating accurate segmentation. We begin with the description of datasets.

Two real-world data sets were used in the experiments and detailed statistics are summarized in Table 2.4.

Table 2.4: Statistics about datasets

Dataset	#Docs	#Words	#Labels
Academia	2.77M	91.6M	300
Yelp	4.75M	145.1M	300

- The **Academia** dataset[1] is a collection of major computer science journals and proceedings. We use both titles and abstracts in our experiments.

- The **Yelp** dataset[2] provides reviews of 250 businesses. Each individual review is considered as a document.

To demonstrate the effectiveness of the proposed approach, we compared the following phrase extraction methods.

- **TF-IDF** ranks phrases by the product of their raw frequencies and inverse document frequencies.

- **C-Value** proposes a ranking measure based on frequencies of a phrase used as parts of their super-phrases following a top-down scheme.

- **ConExtr** approaches phrase extraction as a market-baskets problem based on an assumption about relationship between n-gram and prefix/suffix $(n\text{-}1)$-gram.

- **KEA**[3] is a supervised keyphrase extraction method for long documents. To apply this method in our setting, we consider the whole corpus as a single document.

- **TopMine**[4] is a topical phrase extraction method. We use its phrase mining module for comparison.

- **ClassPhrase** ranks phrases based on their estimated qualities (removing step 3–5 from our framework).

---

[1] http://aminer.org/billboard/AMinerNetwork
[2] https://www.yelp.com/academic_dataset
[3] https://code.google.com/p/kea-algorithm
[4] http://web.engr.illinois.edu/~elkishk2/

- **SegPhrase** combines ClassPhrase with phrasal segmentation to filter overestimated phrases based on normalized rectified frequency (removing step 4 from our framework).

- **SegPhrase**+ is similar to SegPhrase but adds segmentation features to refine quality estimation. It contains the full procedures presented in Section 2.3.

The first two methods utilize NLP chunking to obtain phrase candidates. We use the JATE[5] implementation of the first two methods, i.e., TF-IDF and C-Value. Both of them rely on OpenNLP[6] as the linguistic processor to detect phrase candidates in the corpus. The rest methods are all based on frequent n-grams and the runtime is dramatically reduced. The last three methods are variations of our proposed method.

It is also worth mentioning that JATE contains several more implemented methods including Weirdness [Ahmad et al., 1999]. They are not reported here due to their unsatisfactory performance compared to the baselines listed above.

For the parameter setting, we set minimum phrase support τ as 30 and maximum phrase length ω as 6, which are two parameters required by all methods. Other parameters required by baselines were set according to the open source tools or the original papers.

For our proposed methods, training labels for phrases were collected by sampling representative phrase candidates from groups of phrases pre-clustered on the normalized feature space by k-means. We labeled research areas, tasks, algorithms and other scientific terms in the Academia dataset as quality phrases. Some examples are "divide and conquer," "np complete," and "relational database." For the Yelp dataset, restaurants, dishes, cities and other related concepts are labeled to be positive. In contrast, phrases like "under certain assumptions," "many restaurants," and "last night" were labeled as negative. We downsample low-quality phrases because they are dominant over quality phrases. The number of training labels in our experiments are reported in Table 2.4. To automatically learn the value of segment length penalty, we set the non-segmented ratio r_0 in Algorithm 5 as 1.0 for Academia dataset and 0.95 for Yelp dataset. The selection of this parameter will be discussed in detail later in this section.

To make outputs returned by different methods comparable, we converted all the phrase candidates to lower case and merged plural with singular phrases. The phrase lists generated by these methods have different size, and the tail of the lists are low quality. For the simplicity of comparison, we discarded low-ranked phrases based on the minimum size among all phrase lists except Conextr. Conextr returns all phrases without ranking. Thus, we did not remove its phrases. The remaining size of each list is still reasonably large ($> 40,000$).

2.4.1 QUANTITATIVE EVALUATION AND RESULTS

The goal of our experiments is to study how well our methods perform in terms of "precision" and "recall" and compare with baselines. Precision is defined as the ratio of "true" quality phrases

---

[5]https://code.google.com/p/jatetoolkit
[6]http://opennlp.apache.org

among all predictions. Recall is defined as the ratio between "true" quality phrases in the predictions and the complete set of quality phrases.

Wiki Phrases: The first set of experiments were conducted by using Wikipedia phrases as ground truth labels. Wiki phrases refer to popular mentions of entities by crawling intra-Wiki citations within Wiki content. To compute precision, only the Wiki phrases are considered to be positive. For recall, we combine Wiki phrases returned by different methods altogether and view them as all quality phrases. Precision and recall are biased in this case because positive labels are restricted to Wiki phrases. However, we still expect to obtain meaningful insights regarding the performance difference between the proposed and baselines.

Pooling: Besides Wiki phrases, we rely on human evaluators to judge whether the rest of the candidates are good. We randomly sampled k Wiki-uncovered phrases from the returned candidates of each compared method. These sampled phrases formed a pool and each of them was then evaluated by three reviewers independently. The reviewers could use a popular search engine for the candidates (thus helping reviewers judge the quality of phrases that they were not familiar with). We took the majority of the opinions and used these results to evaluate the methods on how precise the returned quality phrases are. Throughout the experiments we set $k = 500$.

Precision-recall curves of different methods evaluated by both Wiki phrases and pooling phrases are shown in Figure 2.2. The trends on both datasets are similar.

Among the existing work, the chunking-based methods, such as TF-IDF and C-Value, have the best performance; Conextr reduces to a dot in the figure since its output does not provide the ranking information. Our proposed method, SegPhrase+, outperforms them significantly. More specifically, SegPhrase+ can achieve a higher recall while its precision is maintained at a satisfactory level. That is, many more quality phrases can be found by SegPhrase+ than baselines. Under a given recall, precision of our method is higher in most of the time.

For variant methods within our framework, it is surprising that ClassPhrase could perform competitively to the chunking-based methods like TF-IDF. Note that the latter requires large amounts of pre-training for good phrase chunking. However, ClassPhrase's precision at the tail is slightly worse than TF-IDF on Academia dataset evaluated by Wiki phrases. We also observe a significant difference between SegPhrase and ClassPhrase, indicating phrasal segmentation plays a crucial role to address the completeness criterion. In fact, SegPhrase already beats ClassPhrase and baselines. Moreover, SegPhrase+ improves the performance of SegPhrase, because of the use of phrasal segmentation results as additional features.

An interesting observation is that the advantage of our method is more significant on the pooling evaluations. The phrases in the pool are not covered by Wiki, indicating that Wikipedia is not a complete source of quality phrases. However, our proposed methods, including SegPhrase+, SegPhrase, and ClassPhrase, can mine out most of them (more than 80%) and keep a very high level of precision, especially on the Academia dataset. Therefore, the evaluation results on the

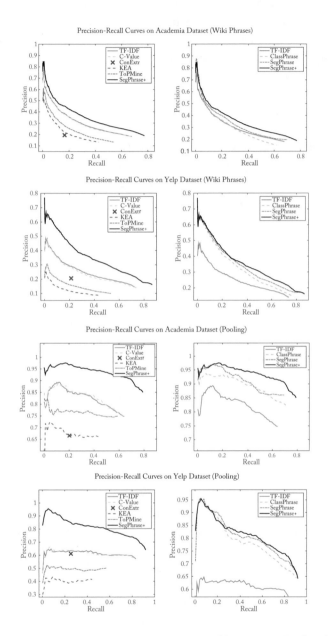

Figure 2.2: Precision-recall in four groups of experiments: (Academia, Yelp) × (Wiki phrase, pooling).

pooling phrases suggest that our methods not only detect the well-known Wiki phrases, but also work properly for the long tail phrases which might occur not so frequently.

From the result on Yelp dataset evaluated by pooling phrases, we notice that SegPhrase+ is a little weaker than SegPhrase at the head. As we know, SegPhrase+ has tried to utilize phrasal segmentation results from SegPhrase to refine the phrase quality estimator. However, segmentation features do not add new information for bigrams. If there are not many quality phrases with more than two words, SegPhrase+ might not have significant improvement and even can perform slightly worse due to the overfitting problem by reusing the same set of labeled phrases. In fact, on Academia dataset, the ratios of quality phrases with more than 2 words are 24% among all Wiki phrases and 17% among pooling phrases. In contrast, these statistics go down to to 13% and 10% on Yelp dataset, which verifies our conjecture and explains why SegPhrase+ has slightly lower precision than SegPhrase at the head.

2.4.2   MODEL SELECTION

The goal of model selection is to study how well our methods perform in terms of "precision" and "recall" on various candidate models with different parameters. We specifically want to study four potentially interesting questions.

- How many labels do we need to achieve good results of phrase quality estimation?

- How to choose non-segmented ratio r_0 for deciding segment length penalty?

- How many iterations are needed to alternate between phrase quality estimation and phrasal segmentation?

- What is the convergence speed of viterbi training?

Number of Labels

To evaluate the impact of training data size on the phrase quality estimation, we focus on studying the classification performance of ClassPhrase. Table 2.5 shows the results evaluated among phrases with positive predictions (i.e., $\{v \in \mathcal{P} : Q_v \geq 0.5\}$. With different numbers of labels, we report the precision, recall and F1 score judged by human evaluators (Pooling). The number of correctly predicted Wiki phrases is also provided together with the total number of positive phrases predicted by the classifier. From these results, we observe that the performance of the classifier becomes better as the number of labels increases. Specifically, on both datasets, the recall rises up as the number of labels increases, while the precision goes down. The reason is the downsampling of low-quality phrases in the training data. Overall, the F1 score is monotonically increasing, which indicates that more labels may result in better performance. 300 labels are enough to train a satisfactory classifier.

Table 2.5: Impact of training data size on ClassPhrase (Top: Academia, Bottom: Yelp)

Academia					
# Labels	Precision	Recall	F1	# Wiki Phrases	# Total
50	0.881	0.372	0.523	6,179	24,603
100	0.859	0.430	0.573	6,834	30,234
200	0.856	0.558	0.676	8,196	40,355
300	0.760	0.811	0.785	11,535	95,070
Yelp					
# Labels	Precision	Recall	F1	# Wiki Phrases	# Total
50	0.491	0.948	0.647	6,985	79,091
100	0.540	0.948	0.688	6,692	57,018
200	0.554	0.948	0.700	6,786	53,613
300	0.559	0.944	0.702	6,777	53,442

Non-segmented Ratio

The non-segmented ratio r_0 is designed for learning segment length penalty, which further controls the precision and recall phrasal segmentation. Empirically, under higher r_0, the segmentation process will favor longer phrases, and vice versa. We show experimental results in Table 2.6 for models with different values of r_0. The evaluation measures are similar to the previous setting but they are computed based on the results of SegPhrase. One can observe that the precision increases with lower r_0, while the recall decreases. It is because phrases are more likely to be segmented into words by lower r_0. High r_0 is generally preferred because we should preserve most positive phrases in training data. We select $r_0 = 1.00$ and 0.95 for Academia and Yelp datasets respectively, because quality phrases are shorter in Yelp dataset than in Academia dataset.

Convergence Study of Viterbi Training

Our time complexity analysis in Section 2.3.5 assumes Viterbi Training in Algorithm 4 converges in few iterations. Here we verify this through empirical studies. From Table 2.7, VT converges extremely fast on both datasets. This owes to the good initialization based on raw frequency as well as the particular property of Viterbi Training discussed in Section 2.3.3.

Iterations of SegPhrase+

SegPhrase+ involves only one iteration of re-estimating phrase quality using normalized rectified frequency from phrasal segmentation. Here we show the performance of SegPhrase+ with more iterations in Figure 2.3 based on human-labeled phrases. For comparison, we also report

Table 2.6: Impact of non-segmented ratio r_0 on SegPhrase (Top: Academia, Bottom: Yelp)

Academia					
r_0	Precision	Recall	F1	# Wiki Phrases	# Total
1.00	0.816	0.756	0.785	10,607	57,668
0.95	0.909	0.625	0.741	9,226	43,554
0.90	0.949	0.457	0.617	7,262	30,550
0.85	0.948	0.422	0.584	7,107	29,826
0.80	0.944	0.364	0.525	6,208	25,374
Yelp					
r_0	Precision	Recall	F1	# Wiki Phrases	# Total
1.00	0.606	0.948	0.739	7,155	48,684
0.95	0.631	0.921	0.749	6,916	42,933
0.90	0.673	0.846	0.749	6,467	34,632
0.85	0.714	0.766	0.739	5,947	28,462
0.80	0.725	0.728	0.727	5,729	26,245

Table 2.7: Objective function values of Viterbi Training for SegPhrase and SegPhrase+

Dataset	Academia		Yelp	
Method	SegPhrase	SegPhrase+	SegPhrase	SegPhrase+
Iter.1	-6.39453E+08	-6.33064E+08	-9.33899E+08	-9.27055E+08
Iter.2	-6.23699E+08	-6.17419E+08	-9.12082E+08	-9.06041E+08
Iter.3	-6.23383E+08	-6.17214E+08	-9.11835E+08	-9.05946E+08
Iter.4	-6.23354E+08	-6.17196E+08	-9.11819E+08	-9.05940E+08
Iter.5	-6.23351E+08	-6.17195E+08	-9.11818E+08	-9.05940E+08

performance of ClassPhrase+ which is similar with ClassPhrase but contains segmentation feature generated by results of phrasal segmentation from the last iteration.

We can see that the benefits brought by rectified frequency can be fully digested within the first iteration, leaving F1 scores over the next several iterations close. One can also observe a slight performance decline over the next two iterations especially for the top-1000 phrases. Recall that we are reusing training labels for each iteration. Then this decline can be well explained by overfitting because segmentation features added by later iterations become less meaningful. Meanwhile, more meaningless features will undermine the classification power of random forest. Based on this, we can conclude that there is no need to do the phrase quality re-estimation multiple times.

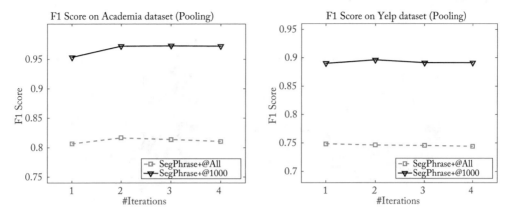

Figure 2.3: Performance variations of SegPhrase+ and ClassPhrase+ with increasing iterations.

2.4.3   EFFICIENCY STUDY

The following execution time experiments were all conducted on a machine with two Intel(R) Xeon(R) CPU E5-2680 v2 @ 2.80 GHz. Our framework is mainly implemented in C++ while a small part of preprocessing is in Python. As shown in Figure 2.4, the linear curves of total runtime of SegPhrase+ on different proportions of data verifies our linear time complexity analyzed in Section 2.3.5.

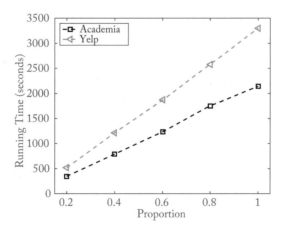

Figure 2.4: Runtime on different proportions of data.

Besides, the pies in Figure 2.5 show the ratios of different components of our framework. One can observe that Feature Extraction and Phrasal Segmentation occupy most of the runtime.

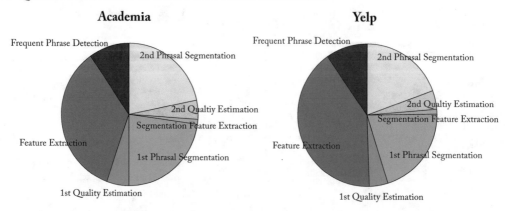

Figure 2.5: Runtime of different modules in our framework on Academia and Yelp dataset.

Fortunately, almost all components of our frameworks can be parallelized, such as Feature Extraction, Phrasal Segmentation, and Quality Estimation, which are the most expensive parts of execution time. It is because sentences can be proceeded one by one without any impact on each other. Therefore, our methods could be very efficient for massive corpus using parallel and distributed techniques. Here we do not compare the runtime with other baselines because they are implemented by different programming languages and some of them further rely on various third-party packages. Among existing implementations, our method is empirically one of the fastest.

2.4.4   CASE STUDY

Previous experiments are focused on evaluating phrase quality quantitatively. In this subsection, we show two case studies based on applications taking segmented corpora as input. Note that the segmented corpus can be obtained by applying the segmenter (i.e., the other output of the phrase mining methods) to the training corpus.

Interesting Phrase Mining
The first application is to mine interesting phrases in a subset of given corpus. Interesting phrases are defined to be phrases frequent in the subset C' but relatively infrequent in the overall corpus

Table 2.8: Running time

Dataset	File Size	# Words	Time
Academia	613 MB	91.6 M	0.595 h
Yelp	750 MB	145.1 M	0.917 h

$\mathcal{C}$ [Bedathur et al., 2010, Gao and Michel, 2012, P et al., 2014]. Given a phrase v, its interesting-ness is measured by $freq(v, \mathcal{C}') \cdot purity(v, \mathcal{C}', \mathcal{C}) = freq(v, \mathcal{C}')^2/freq(v, \mathcal{C})$, which considers both phrase frequency and purity in the subset.

We list a fraction of interesting phrases in Table 2.9 mined from papers published in SIGMOD and SIGKDD conferences. Each series of proceedings form a subset of the whole Academia corpus. Two segmentation methods are compared. The first one relies on dynamic programming using phrase quality estimated by SegPhrase+. The other is based on the phrase chunking method adopted in JATE, which is further used to detect phrase candidates for TF-IDF and C-Value methods. To be fair, we only show phrases extracted by SegPhrase+, TF-IDF, and C-Value methods in the table. Because TF-IDF and C-Value perform similarly and they both rely on the chunking method, we merge their phrases and report mining results in one column named "Chunking." Phrases in SegPhrase+ but missing in the chunking results are highlighted in purple (red vice versa). One can observe that the interesting phrases mined by SegPhrase+ based on the segmentation result are more meaningful and the improvement is significant. Relatively speaking, phrases mined from the chunking method are of inferior quality. Therefore, many of them are not covered by SegPhrase+.

Word/Phrase Similarity Search

With a segmented corpus, one could train a model to learn *distributed vector representations* of words and phrases [Mikolov et al., 2013]. Using this technique, words and phrases are mapped into a vector space such that semantically similar words and phrases have similar vector represen-tations. It helps other text mining algorithms to achieve better performance by grouping similar units. The quality of the learned vector representation is closely related to the quality of the input segmented corpus. Accurate segmentation results in good vector representation and this perfor-mance gain is usually evaluated by comparing similarity scores between word/phrase pairs. To be specific, one could compute top-k similar words or phrases given a query and compare the ranked lists. We use this to verify the utility of both quality phrase mining and quality segmentation.

We show the results in Table 2.10 from SegPhrase+ and the chunking method mentioned in the previous interesting phrase mining application. Queries were chosen to be capable of show-ing the difference between the two methods for both Academia and Yelp datasets. Distributed representations were learned through an existing tool [Mikolov et al., 2013] and ranking scores were computed based on cosine similarity.

From the table, one can easily tell that the rank list from SegPhrase+ carries more sense than that from phrase chunking. One of the possible reasons is that chunking method only detects noun phrases in the corpus, providing less accurate information of phrase occurrences than SegPhrase+ to the vector representation learning algorithm.

Table 2.9: Interesting phrases mined from papers published in SIGMOD and SIGKDD

	SIGMOD		SIGKDD	
	SegPhrase+	Chunking	SegPhrase+	Chunking
1	data base	data base	data mining	data mining
2	database system	database system	data set	association rule
3	relational database	query processing	association rule	knowledge discovery
4	query optimization	query optimization	knowledge discovery	frequent itemset
5	query processing	relational database	time series	decision tree
...	...	...	...	...
51	sql server	database technology	assoc. rule mining	search space
52	relational data	database server	rule set	domain knowledge
53	data structure	large volume	concept drift	important problem
54	join query	performance study	knowledge acquisition	concurrency control
55	web service	web service	gene expression data	conceptual graph
...	...	...	...	...
201	high dimensio. data	efficient impl.	web content	optimal solution
202	location based serv.	sensor network	frequent subgraph	semantic relation
203	xml schema	large collection	intrusion detection	effective way
204	two phase locking	important issue	categorical attribute	space complexity
205	deep web	frequent itemset	user preference	small set
...	...	...	...	...

Case Study of Quality Phrases

We show some phrases from ranking lists generated by ClassPhrase, SegPhrase, and SegPhrase+ in Table 2.11. In general, phrase quality drops with number goes up. ClassPhrase always performs the worst among the three methods. SegPhrase+ is slightly better than SegPhrase, which is noticeable for phrases ranked after 20,000. It's worth mentioning that the minimum sizes of phrase lists are 50,577 and 42,989 for two datasets, respectively.

2.5   SUMMARY

In this chapter, we introduced a data-driven model for extracting quality phrases from text corpora with user guidance. By requiring limited training effort, the model can achieve outstanding performance even for highly irregular textual datausiness reviews. The key idea is to rectify the raw frequency of phrases which misleads quality estimation. A segmentation-integrated approach is

Table 2.10: Top-5 similar phrases for representative queries (Top: Academia, Bottom: Yelp)

Query	Data Mining		Olap	
Method	SegPhrase+	Chunking	SegPhrase+	Chunking
1	knowledge discovery	driven methodologies	data warehouse	warehouses
2	text mining	text mining	online analy. proc.	clustcube
3	web mining	financial investment	data cube	rolap
4	machine learning	knowledge discovery	olap queries	online analy. proc.
5	data mining techniques	building knowledge	multidim. databases	analytical processing

Query	Blu-ray		Noodle		Valet Parking	
Method	Seg-Phrase+	Chunking	Seg-Phrase+	Chunking	Seg-Phrase+	Chunking
1	dvd	microwave	ramen	noodle soup	valet	huge lot
2	vhs	lifetime wty	noodle soup	asian noodle	self-parking	private lot
3	cd	recliner	rice noodle	beef noodle	valet service	self-parking
4	new release	battery	egg noodle	stir fry	free valet parking	valet
5	sony	new battery	pasta	fish ball	covered parking	front lot

therefore developed and finally addresses such a fundamental limitation of phrase mining. However, we discover that despite the outstanding performance, the reliance on manual efforts from domain experts can still become an impediment for timely analysis of massive, emerging text corpora. A fully automated algorithm, instead, can be much more useful in this scenario. Meanwhile, this chapter focuses on multi-word phrase mining while single-word phrases are not taken care of. The integration of light-weight linguistic processors such as POS tagging is also worth studying. We reserve these topics for the next chapter.

Table 2.11: Sampled quality phrases from Academia and Yelp datasets *(Continues.)*

Academia			
Method	ClassPhrase	SegPhrase	SegPhrase+
1	virtual reality	virtual reality	self organization
2	variable bit rate	variable bit rate	polynomial time approx.
3	shortest path	shortest path	least squares
...	...	...	...
501	finite state	frequency offset estimation	health care
502	air traffic	collaborative filtering	gene expresion
503	long term	ultra wide band	finite state transducers
...	...	...	...
2001	chemical reaction	ad hoc networks	quasi monte carlo
2002	container terminals	hyperspectral remote sensing	integer programming
2003	graceful degradation	piecewise affine	gray level
...	...	...	...
10001	search terms	test plan	airline crew scheduling
10002	high dimensional space	automatic text	integer programming
10003	delay variation	adaptive bandwidth	web log data
...	...	...	...
20001	test coverage	implementation costs	experience sampling
20002	adaptive sliding mode control	error bounded	virtual execution environments
20003	random graph models	free market	nonlinear time delay systems
...	...	...	...
50001	svm method	harmony search algorithm	asymptotic theory
50002	interface adaptation	integer variables	physical mapping
50003	diagnostic fault simulation	nonlinear oscillators	distince patterns
...	...	...	...

Table 2.11: *(Continued.)* Sampled quality phrases from Academia and Yelp datasets

Yelp			
Method	ClassPhrase	SegPhrase	SegPhrase+
1	taco bell	taco bell	tour guide
2	wet republic	wet republic	yellow tail
3	pizzeria bianco	pizzeria bianco	vanilla bean
...	...	...	...
501	panoramic view	art museum	rm seafood
502	pretzel bun	ice cream parlor	pecan pie
503	spa pedicure	pho kim long	master bedroom
...	...	...	...
2001	buffalo chicken wrap	training sessions	smashed potatoes
2002	salvation army	folding chairs	italian stallion
2003	shortbread cookies	single bypass	ferris wheel
...	...	...	...
10001	seated promptly	carrot soup	gary danko
10002	leisurely stroll	veggie soup	benny benassi
10003	flavored water	pork burrito	big eaters
...	...	...	...
20001	buttery toast	late night specials	cilantro hummus
20002	quick breakfast	older women	lv convention center
20003	slightly higher	worth noting	iced vanilla
...	...	...	...
40001	friday morning	conveniently placed	coupled with
40002	start feeling	cant remember	way too high
40003	immediately start	stereo system	almost guaranteed
...	...	...	...

CHAPTER 3

Automated Quality Phrase Mining

Almost all state-of-the-art methods in NLP, IR, and text mining communities require human experts at certain levels. For example, NLP-based methods [Frantzi et al., 2000, Park et al., 2002, Zhang et al., 2008] require language experts for *rigorous language rules* or *sophisticated labels* (e.g., parsing tree labels) to identify phrase mentions. SegPhrase+ introduced in last chapter doesn't rely on linguistic processing and outperforms many other methods [Deane, 2005, El-Kishky et al., 2015, Frantzi et al., 2000, Parameswaran et al., 2010, Park et al., 2002, Zhang et al., 2008], but needs *hundreds of binary labels* telling whether a phrase is of high quality. Such reliance on manual efforts from domain experts becomes an impediment for timely analysis of massive, emerging text corpora. Besides this issue, an ideal *automated phrase mining* method, as shown in Figure 3.1, is supposed to work smoothly for multiple languages with high performance in terms of precision, recall, and efficiency.

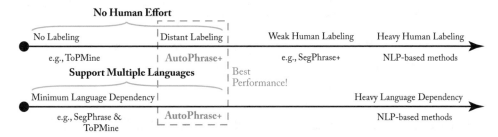

Figure 3.1: Motivation: Automated phrase mining without human effort for multiple languages.

3.1 OVERVIEW

Toward the goal of making the framework fully automated, we summarize the following three major challenges.

1. *Can we completely remove the human effort for labeling phrases?* In the previous chapter, SegPhrase+ has shown that the quality of phrases generated by unsupervised methods [Deane, 2005, El-Kishky et al., 2015, Parameswaran et al., 2010] is acceptable but

much weaker than the supervised methods, and at least a few hundred labels are necessary for training. *Distant training* is a popular methodology to reduce expensive human labor by utilizing high-quality phrases in knowledge bases as positive phrase labels.

2. *Can we achieve high performance of phrase mining in multiple languages?* Complicated pre-processing models, such as dependency parsing, heavily rely on human efforts and thus cannot be smoothly applied to multiple languages, as shown in Figure 3.1. To achieve high performance with minimum language dependency, we fully utilize the results of the following two techniques: (1) *tokenization* should be allowed because it provides the building bricks of phrases—the boundaries of words; and (2) *part-of-speech (POS) tagging*, another elementary preprocessing step in NLP pipelines, is available in most of languages. And there are language-independent part-of-speech taggers, such as TreeTagger [Schmid, 2013]. Moreover, Observation 3.1 suggests that the context information from POS tags can be a strong signal for identifying the phrase boundary in complement to the frequency-based statistical signals.

Observation 3.1 Combining frequency-based signals with POS information is helpful.

#1	[Sophia	Smith]	was	born	in	England	.
	NNP	NNP	VBD	VBN	IN	NNP	.
#2	...	the	[Great	Firewall]	is	...	
	...	DT	NNP	NNP	VBZ	...	
#3	This	is	a	great	[firewall	software]	.
	DT	VBZ	DT	JJ	NN	NN	.

The data-driven methods usually rely on frequency-based signals [Deane, 2005, El-Kishky et al., 2015, Liu et al., 2015, Parameswaran et al., 2010] and can lead to two types of errors.

(1) Over-decomposition: Combination of individual popular words tend to be more decomposable. In *#1*, since both *Sophia* and *Smith* are very popular names, to make the full name a complete phrase, *Sophia Smith* is required to be also popular, which may not be true.

(2) Under-decomposition: Popular phrases tend to be less decomposable. For instance, "*great firewall*" are mentioned in both *#2* and *#3*. Suppose this phrase is mentioned frequently in our corpus, it may prevent the algorithm to extract "*firewall system*" from *#3* if it is less popular. However, POS information can be helpful to avoid such faults. In *#1*, two consecutive nouns emit a strong indicator for a phrase; in *#2*, the transition from the noun "*Firewall*" to the verb "*is*" implies a phrase boundary; in *#3*, "*great*" is an adjective while both "*firewall*" and "*software*" are nouns, making "*firewall software*" a more likely unit.

#4	*The*	[*discriminative* *classifier*]	[*SVM*]	*is*	...
	DT	*JJ* NN	*NN*	*VBZ*	...

On the other hand, purely considering POS tags may not be wise regardless of the tagging performance. For example, in *#4*, "*classifier SVM*" will be wrongly extracted if only POS tags are considered. In this case, frequency-based signals can correct the error. □

3. *Can we simultaneously model single-word and multi-word phrases?* In linguistic analysis, a phrase is not only a group of multiple words, but also possibly a single word, as long as it functions as a constituent in the syntax of a sentence [Finch, 2000]. As a great portion (ranging from 10–30% based on our experiments) of high-quality phrases, we should take single-word phrases (e.g., ⌈ UIUC ⌋, ⌈ Illinois ⌋, and ⌈ USA ⌋) into consideration as well as multi-word phrases to achieve a high recall in phrase mining.

In this chapter, we introduce a novel automated phrase mining method AutoPhrase+ to address these three challenges simultaneously, mainly using the following three techniques.

1. *Positive-only Distant Training.* Distant training is utilized to generate clean positive but noisy negative labels from a knowledge base for ensemble classifiers to estimate phrase quality scores. Thereafter, human labeling is no longer needed.

2. *POS-guided phrasal segmentation.* This technique utilizes the context information embedded in POS tags and accurately locates the boundaries of phrases in the given corpus, which improves precision.

3. *Single-word phrase modeling.* The mining framework designed for multi-word phrase mining is extended for single-word phrases and gains about 10–30% more recall.

In addition, the language dependency is minimized: AutoPhrase+ only requires tokenization and POS tagging in the preprocessing. Theoretically, it is compatible with any language as long as a knowledge base (e.g., Wikipedia), a tokenizer, and a POS tagger in that language are available. Moreover, as demonstrated in our experiments, AutoPhrase+ supports English, Spanish and Chinese. To our best knowledge, this is the first phrase mining method that can smoothly work for multiple languages. More importantly, it is adaptable to other languages with minimal engineering cost.

3.2   AUTOMATED PHRASE MINING FRAMEWORK

Figure 3.2 presents the automated phrase mining framework. Different from previous phrase mining approach SegPhrase+ which requires human-generated labels, this new framework takes a knowledge base as the side input. After preprocessing with third party tools including tokenizers from Lucene and Stanford NLP as well as the POS taggers from TreeTagger, AutoPhrase+ includes five modules: frequent phrase mining, noisy label generation, robust positive-only distant training, POS-guided phrasal segmentation, and phrase quality re-estimation.

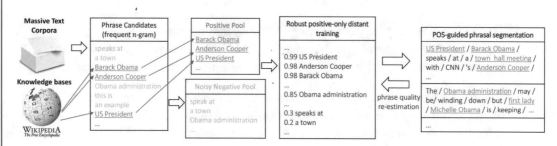

Figure 3.2: The automated phrase mining framework.

3.2.1 PHRASE LABEL GENERATION

To assign quality score to each candidate phrase, as introduced in the previous chapter, SegPhrase+ required domain experts to first carefully select hundreds of varying-quality phrases from millions of candidates, and then annotate them with binary labels. For example, for computer science papers, our domain experts provided hundreds of positive labels (e.g., "spanning tree" and "computer science") and negative labels (e.g., "paper focuses" and "important form of"). However, creating such a label set is expensive, especially in specialized domains like clinical reports and business reviews, because this approach provides no clues for how to identify the phrase candidates to be labeled. In this section, we introduce a method that only utilizes existing general knowledge bases without any other human effort.

Public knowledge bases (e.g., Wikipedia) usually encode a considerable number of high-quality phrases in the titles, keywords, and internal links of pages. For example, by analyzing the internal links and synonyms[1] in English Wikipedia, more than a hundred thousand high-quality phrases were discovered. As a result, we place these phrases in a ***positive pool***.

Knowledge bases, however, rarely, if ever, identify phrases that fail to meet our criteria, what we call *inferior phrases*. An important observation is that the number of phrase candidates, based on *n-grams* (recall leftmost box of Figure 3.2), is huge and the majority of them are actually of of inferior quality (e.g., "speaks at"). In practice, based on our experiments, among millions of phrase candidates, usually, only about 10% are in good quality. Therefore, phrase candidates that are derived from the given corpus but that fail to match any high-quality phrase derived from the given knowledge base, are used to populate a large but noisy ***negative pool***.

Directly training a classifier based on the noisy label pools is not a wise choice: some phrases of high quality from the given corpus may have been missed (i.e., inaccurately binned into the

---

[1]https://github.com/kno10/WikipediaEntities

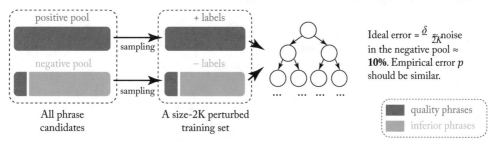

Figure 3.3: The illustration of each base classifier. In each base classifier, we first randomly sample K positive and negative labels from the pools respectively. There might be δ quality phrases among the K negative labels. An unpruned decision tree is trained based on this perturbed training set.

negative pool) simply because they were not present in the knowledge base. Instead, we propose to utilize an ensemble classifier that averages the results of T independently trained base classifiers. As shown in Figure 3.3, for each base classifier, we randomly draw K phrase candidates with replacement from the positive pool and the negative pool respectively (considering a canonical balanced classification scenario). This size-$2K$ subset of the full set of all phrase candidates is called a ***perturbed training set*** Breiman [2000], because the labels of some (δ in the figure) quality phrases are switched from positive to negative. In order for the ensemble classifier to alleviate the effect of such noise, we need to use base classifiers with the lowest possible training errors. We grow an unpruned decision tree to the point of separating all phrases to meet this requirement. In fact, such decision tree will always reach 100% training accuracy when no two positive and negative phrases share identical feature values in the perturbed training set. In this case, its ideal error is $\frac{\delta}{2K}$, which approximately equals to the proportion of switched labels among all phrase candidates (i.e., $\frac{\delta}{2K} \approx 10\%$). Therefore, the value of K is not sensitive to the accuracy of the unpruned decision tree and is fixed as 100 in our implementation. Assuming the extracted features are distinguishable between quality and inferior phrases, the empirical error evaluated on all phrase candidates, p, should be relatively small as well.

An interesting property of this sampling procedure is that the random selection of phrase candidates for building perturbed training sets creates classifiers that have statistically independent errors and similar erring probability Breiman [2000], Martínez-Muñoz, and Suárez [2005]. Therefore, we naturally adopt random forest Geurts, Ernst and Wehenkel [2006], which is verified, in the statistics literature, to be robust and efficient. The phrase quality score of a particular phrase is computed as the proportion of all decision trees that predict that phrase is a quality phrase. Suppose there are T trees in the random forest, the ensemble error can be estimated as the probability of having more than half of the classifiers misclassifying a given phrase candidate as follows.

$$\text{ensemble\_ error}(T) = \sum_{t=\lfloor 1+T/2 \rfloor}^{T} \cdot \binom{T}{t} p^t (1-p)^{T-t}.$$

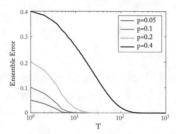

Figure 3.4: Ensemble errors of different p's varying T.

From Figure 3.4, one can easily observe that the ensemble error is approaching 0 when T grows. In practice, T needs to be set larger due to the additional error brought by model bias. Empirical studies can be found in Figure 3.8.

3.2.2 PHRASE QUALITY ESTIMATION

In the last chapter, we have introduced in detail about the four criteria for measuring multi-word phrase quality, i.e., popularity, concordance, informativeness, and completeness. Are they also applicable for single-word phrases? Not necessarily.

Because **single-word phrases** cannot be decomposed into two or more parts, the **concordance** is no longer definable. As the complement, we propose the **independence** requirement for **quality single word phrases** as below.

Independence.   A quality single-word phrase is more likely a complete semantic unit in the given documents. For example, "*UIUC*" is a quality single-word phrase. However, "*united,*" usually occurring within other quality multi-word phrases such as "*United States,*" "*United Kingdom,*" "*United Airlines,*" and "*United Parcel Service,*" is not a quality single-word phrase, because its independence is not enough.

Informativeness Features.   In information retrieval, stop words and inverse document frequency (IDF) are two useful approaches to measure the word informativeness:

- *Stop word*. Whether this word is a stop word; and

- *IDF* of this word.

In general, quality single-word phrases are expected to be a non-stop word with relatively large IDF.

Punctuation is commonly appearing across different languages, especially *quotes*, *brackets*, and *capitalization*. Therefore, we adopt (1) the probability that a single-word phrase is surrounded by quotes or brackets and (2) the probability that the first character of a single-word phrase is in *uppercase*. Higher probability usually indicates a single-word phrase being more informative. A good example is in *support vector machines (SVM)*. Note that, in some languages, such as Chinese, these is no uppercase feature.

The features for multi-word phrases in the previous chapter are inherited, including **concordance features** such as *pointwise mutual information* and *pointwise Kullback-Leibler divergence* after decomposing the phrase into two parts and **informativeness features** involving *IDF*, *stop word*, and *punctuation*.

In addition, we propose two new **context-independent completeness features** inspired by Parameswaran et al. [2010]: (1) the ratio between the phrase frequency and the minimum frequency among its *sub-phrases*; and (2) the ratio between the maximum frequency among its *super-phrases* and the phrase frequency. A low *sub-phrase* ratio usually indicates the phrase can be shorten, while a high *super-phrase* ratio implies the phrase is not complete. For instance, *"NP-complete in the strong"* tends to have a high *super-phrase* ratio because it always occurs in *"NP-complete in the strong sense;"* *"classifier SVM"* is expected to receive a low *sub-phrase* ratio because both *"classifier"* and *"SVM"* are popular elsewhere.

3.2.3   POS-GUIDED PHRASAL SEGMENTATION

POS-guided phrasal segmentation, which is the most crucial component in AutoPhrase+, is proposed to tackle the challenge of measuring **completeness** and **independence** through locating every phrase mention in the corpus and rectifying phrase mentions previously obtained via string match.

Definition 3.2   POS-guided Phrasal Segmentation.   The *"POS-guided"* emphasizes combining with POS tags, which is helpful as indicated in Observation 3.1. Given a corpus $\mathcal{C}$ (i.e., a length-n POS tagged word sequence $\langle w_1 w_2 \ldots w_n, t_1 t_2 \ldots t_n \rangle$), a segmentation $S = s_1 s_2 \ldots s_m$ is induced by a boundary index sequence $B = \{b_1, b_2, \ldots, b_{m+1}\}$ satisfying $1 = b_1 < b_2 < \ldots < b_{m+1} = n+1$, where the i-th segment $s_i = \langle w_{b_i} w_{b_i+1} \ldots w_{b_i+|s_i|-1}, t_{b_i} t_{b_i+1} \ldots t_{b_i+|s_i|-1} \rangle$. Here $|s_i|$ refers to the number of words/tags in segment s_i. Since $b_i + |s_i| = b_{i+1}$, for clarity we use $w_{[b_i,b_{i+1})}$ to denote word sequence $w_{b_i} w_{b_i+1} \cdots w_{b_i+|s_i|-1}$ and $t_{[b_i,b_{i+1})}$ to denote POS tag sequence $t_{b_i} t_{b_i+1} \cdots t_{b_i+|s_i|-1}$. Therefore, the i-th segment $s_i = \langle w_{[b_i,b_{i+1})}, t_{[b_i,b_{i+1})} \rangle$. $\square$

For a better understanding of the POS-guided phrasal segmentation, we provide the following example.

Example 3.3   Recall the example sentences in Observation 3.1. Ideal POS-guided phrasal segmentation results are as follows.

#1:	⟨Sophia Smith, NNP NNP⟩, ⟨was VBD⟩, ⟨born, VBN⟩, ⟨in, IN⟩, ⟨England, NNP⟩, ⟨., .⟩
#2:	..., ⟨the, DT⟩, ⟨Great Firewall, NNP NNP⟩, ⟨is, VBZ⟩ ...
#3:	⟨This, DT⟩, ⟨is, VBZ⟩, ⟨a, DT⟩, ⟨great, JJ⟩, ⟨firewall software, NN NN, ., .
#4	⟨The, DT⟩, ⟨discriminative classifier, JJ NN⟩, ⟨SVM, NN⟩, ⟨is, VBZ⟩, ...

Definition 3.4 POS Sequence Quality.   is defined to be the probability of a word sequence being a complete semantic unit given its corresponding POS tag sequence, according to the above criteria. Given a length-k POS tag sequence $t_1 t_2 \ldots t_k$, its POS sequence quality is:

$$T(t_1 \ldots t_k) = p(\lceil v_1 \ldots v_k \rceil | \text{tag}(v_1 \ldots v_k) = t_1 \ldots t_k) \in [0, 1],$$

where $\text{tag}(v_1 \ldots v_k)$ is the corresponding POS tag sequence of the word sequence $v_1 \ldots v_k$. □

The estimator for POS sequence quality will also be learned, which is expected to work as follows.

Example 3.5   A good POS sequence quality estimator can return $T(\text{NN NN}) \approx 1$, $T(\text{NN VB}) \approx 0$, and $T(\text{DT NN}) \approx 0$, where NN refers to singular or mass noun (e.g., database), VB means verb in the base form (e.g., is), and DT is for determiner (e.g., the).

The POS sequence quality score $T(\cdot)$ is designed to reward the phrases with meaningful POS patterns. The particular form we chosen is:

$$T(t_{[b_i, b_{i+1})}) = \left(1 - \delta(t_{b_{i+1}-1}, t_{b_{i+1}})\right) \times \prod_{j=b_i+1}^{b_{i+1}-1} \delta(t_{j-1}, t_j),$$

where $\delta(t_1, t_2)$ is the probability that the POS tag t_2 is exactly after the POS tag t_1 within a phrase in the given document collection. In this formula, the first term represents that there is a phrase boundary between $b_{i+1} - 1$ and b_i, while the product indicates that all POS tags among $t_{[b_i, b_{i+1})}$ are in the same phrase. This POS quality score can naturally counter the bias to longer segments because exactly one of $\delta(t_1, t_2)$ and $(1 - \delta(t_1, t_2))$ is always multiplied no matter how the corpus is segmented. Note that the length penalty model in SegPhrase+ is a special case when $\delta(t_1, t_2)$ shares the same corresponding value.

Mathematically, $\delta(t_1, t_2)$ is defined as:

$$\delta(t_1, t_2) = p(\ldots \lceil \ldots w_1 w_2 \ldots \rceil \ldots | \mathcal{C}, \text{tag}(w_1) = t_1 \wedge \text{tag}(w_2) = t_2).$$

As it depends on how documents are segmented into phrases, $\delta(t_1, t_2)$ will be learned during the context-aware phrasal segmentation.

Now, after we have both phrase quality Q and POS sequence quality T ready, we are able to formally define the POS-guided phrasal segmentation model. The joint probability of a corpus $\mathcal{C}$ and a segmentation $S = s_1 \ldots s_m$ is factorized as:

$$p(S, \mathcal{C}) = \prod_{i=1}^{m} p\left(b_{i+1}, \lceil w_{[b_i, b_{i+1})} \rfloor \big| b_i, t_{[b_i, b_{i+1})}\right),$$

where $p(b_{i+1}, \lceil w_{[b_i, b_{i+1})} \rfloor | b_i, t_{[b_i, b_{i+1})})$ is the probability of observing a word sequence $w_{[b_i, b_{i+1})}$ as the i-th quality segment given the previous boundary index b_i and its corresponding POS tag sequence $t_{[b_i, b_{i+1})}$.

Since the phrase segments function as a constituent in the syntax of a sentence [Finch, 2000], they usually have weak dependence on each other. As a result, we assume these segments in the word sequence are generated one by one for the sake of both efficiency and simplicity.

For each segment, given the POS tag sequence t and the start index b_i of a segment s_i, the generative process is defined as follows.

1. Generate the end index b_{i+1}, according to its POS sequence quality

$$p(b_{i+1}|b_i, t_{[b_i, b_{i+1})}) = p(\lceil w \rfloor | t_{[b_i, b_{i+1})}) = T(t_{[b_i, b_{i+1})}).$$

2. Given the two ends b_i, b_{i+1}, generate the word sequence $w_{[b_i, b_{i+1})}$ according to a multi-nomial distribution over all segments of length-$(b_{i+1} - b_i)$.

$$p(w_{[b_i, b_{i+1})}|b_i, b_{i+1}) = p\left(w_{[b_i, b_{i+1})}\big||s_i| = b_{i+1} - b_i\right).$$

3. Finally, we generate an indicator whether $w_{[b_i, b_{i+1})}$ forms a quality segment according to its quality

$$p(\lceil w_{[b_i, b_{i+1})} \rfloor | w_{[b_i, b_{i+1})}) = Q(w_{[b_i, b_{i+1})}).$$

Integrating the above three generative steps together, we have the the following probabilistic factorization:

$$
\begin{aligned}
&p(b_{i+1}, \lceil w_{[b_i, b_{i+1})} \rfloor | b_i, t_{[b_i, b_{i+1})}) \\
&= p(b_{i+1}|b_i, t_{[b_i, b_{i+1})}) p(w_{[b_i, b_{i+1})}|b_i, b_{i+1}) p(\lceil w_{[b_i, b_{i+1})} \rfloor | w_{[b_i, b_{i+1})}) \\
&= T(t_{[b_i, b_{i+1})}) p\left(w_{[b_i, b_{i+1})}\big||s_i| = b_{i+1} - b_i\right) Q(w_{[b_i, b_{i+1})}).
\end{aligned}
$$

Therefore, for a given corpus $\mathcal{C}$ with D documents, there are three subproblems:

- learn $p(u||u|)$ for each frequent word and phrase $u \in \mathcal{P}$. We denote $p(u||u|)$ as θ_u for convenience;

Algorithm 6: POS-guided Phrasal Segmentation (PGPS)

---

1 **Input**: Corpus $\mathcal{C} = \langle w_1 w_2 \ldots w_n, t_1 t_2 \ldots t_n \rangle$, phrase quality Q, parameters θ and δ.
2 **Output**: Optimal segmentation S.
    // $h_i \equiv \max_S \quad p(S, \mathcal{C} = \langle w_{[1,i)}, t_{[1,i)} \rangle | Q, \theta, \delta)$
3 $h_1 \leftarrow 1, h_i \leftarrow 0 \ (1 < i \leq n + 1)$
4 **for** $i = 1$ *to* n **do**
5   **for** $j = i + 1$ *to* $n + 1$ **do**
        // Efficiently implemented via Trie.
6   **if** *there is no phrase starting with* $w_{[i,j)}$ **then**
7   **break**
        // In practice, log and addition are used to avoid underflow.
8   **if** $h_i \times p(j, \lceil w_{[i,j)} \rfloor | i, t_{[i,j)}) > h_j$ **then**
9              $h_j \leftarrow h_i \times p(j, \lceil w_{[i,j)} \rfloor | i, t_{[i,j)})$
10             $g_j \leftarrow i$

11 $j \leftarrow n + 1, m \leftarrow 0$
12 **while** $j > 1$ **do**
13     $m \leftarrow m + 1$
14     $s_m \leftarrow \langle w_{[g_j, j)}, t_{[g_j, j)} \rangle$
15     $j \leftarrow g_j$
16 **return** $S \leftarrow s_m s_{m-1} \ldots s_1$

- learn $\delta(t_1, t_2)$ for every POS tag pair; and

- infer the segmentation S when θ and δ are fixed.

We employ the maximum a posterior principle and maximize the joint probability of the corpus:

$$\sum_{d=1}^{D} \log p(S_d, C_d) = \sum_{d=1}^{D} \sum_{i=1}^{m_d} \log p\left(b_{i+1}^{(d)}, \lceil w_{[b_i, b_{i+1})}^{(d)} \rfloor \,\middle|\, b_t^{(d)}, t_{[b_i, b_{i+1})}\right). \tag{3.1}$$

Given the θ and $\delta(\cdot, \cdot)$, to find the best segmentation that maximizes Equation (3.1), we develop an efficient dynamic programming algorithm for the POS-guided phrasal segmentation (PGPS) as shown in Algorithm 6.

When the segmentation S and the parameter θ are fixed, the closed-form solution of $\delta(t_1, t_2)$ is:

$$\delta(t_1, t_2) = \frac{\sum_{d=1}^{D} \sum_{i=1}^{m_d} \sum_{j=b_i^{(d)}}^{b_{i+1}^{(d)}-2} \mathbf{1}(t_j^{(d)} = t_1 \wedge t_{j+1}^{(d)} = t_2)}{\sum_{d=1}^{D} \sum_{i=1}^{n_d-1} \mathbf{1}(t_i^{(d)} = t_1 \wedge t_{i+1}^{(d)} = t_2)}, \tag{3.2}$$

Algorithm 7: AutoPhrase+ Viterbi Training

---

1 **Input**: Corpus $\mathcal{C}$ and phrase quality Q.
2 **Output**: θ and δ.
3 initialize θ with normalized raw frequencies in the corpus
4 **while** θ *does not converge* **do**
5   **while** δ *does not converge* **do**
6   **for** $d = 1$ *to* D **do**
7             $S_d \leftarrow PGPS(C_d, Q, \theta, \delta)$ via Algorithm 6
8   update δ using $S_1 S_2 \ldots S_D$ according to Eq. (3.2)
9   **for** $d = 1$ *to* D **do**
10         $S_d \leftarrow PGPS(C_d, Q, \theta, \delta)$ via Algorithm 6
11   update θ using $S_1 S_2 \ldots S_D$ according to Eq. (3.3)
12 **return** θ and δ

---

where $\mathbf{1}(\cdot)$ denotes the identity indicator and $\delta(t_1, t_2)$ is the unsegmented ratio among all $t_1 t_2$ pairs in the given corpus.

Similarly, once the segmentation S and the parameter δ are fixed, the closed-form solution of θ_u can be derived as:

$$\theta_u = \frac{\sum_{d=1}^{D} \sum_{i=1}^{m_d} \mathbf{1}(w_{[b_i, b_{i+1})}^{(d)} = u)}{\sum_{d=1}^{D} \sum_{i=1}^{m_d} \mathbf{1}(|s_i^{(d)}| = |u|)}. \tag{3.3}$$

We can see that θ_u is the times that u becomes a complete segment normalized by the number of the length-$|u|$ segments.

As shown in Algorithm 7, our optimization strategy for learning δ and θ is a nested iterative optimization process similar to SegPhrase+. In our case, given corpus $\mathcal{C}$, in the inner loop, it first fixes θ and keeps adjusting parameters δ using the segmentation that maximizes $p(\mathcal{S}, \mathcal{C} | Q, \theta, \delta)$ until converge. Later, in the outer loop, δ is fixed and θ will be updated. Such a procedure is iterated until a stationary point has been reached.

As same as that in SegPhrase+, the efficiency is the major reason that we choose Hard-EM instead of finding a maximum likelihood estimator of θ and δ using Soft-EM (i.e., Bawm-Welch algorithm [Bishop, 2006]).

3.2.4 PHRASE QUALITY RE-ESTIMATION

Different from SegPhrase+, instead of adding two more features computed based on the rectified frequency of phrases, we *reconstruct the whole feature space* as follows.

- When generating frequency-related features, such as concordance features and completeness features, the raw frequency is replaced by the rectified frequency.

- When calculating occurrence-related features, such as the informativeness features, only those complete segments matched occurrences are considered.

The reconstruction exploits the rectified frequency in a more thorough way and thus yielding a better performance gain.

In addition, an **independence feature** is added for single-word phrases. Formally, it is the ratio of the rectified frequency of a single-word phrase given the context-aware phrasal segmentation over its raw frequency. Quality single-word phrases are expected to have large values. For example, "*united*" is likely to an almost zero ratio.

3.2.5   COMPLEXITY ANALYSIS

Same as SegPhrase+, the complexity of AutoPhrase+ is theoretically linear to the corpus size and thus very efficient and scalable. Meanwhile, every component can be parallelized in an almost lock-free way grouping by either phrases or sentences. That is, suppose T threads are enabled, the time complexity becomes $O(|\mathcal{C}|/T)$.

3.3   EXPERIMENTAL STUDY

In this section, we will apply the proposed method to mine quality phrases from three large text corpora in three languages (English, Spanish, and Chinese) and validate whether it satisfies the three requirements of automated phrase mining. After introducing the experimental settings, we compare the proposed method with many other methods to demonstrate its high precision and recall as well as the importance of single-word phrase modeling. Then, we explore the robustness of the proposed positive-only distant training and its performance against expert labeling. The importance of incorporating POS tags in the POS-guided phrasal segmentation has also been verified. In the end, we present the case study and the efficiency study.

For the purpose of checking the language dependency of the proposed model, we have prepared three large collections of text in different languages cleaned from **English**, **Spanish**, and **Chinese** Wikipedia articles. Table 3.1 shows the detailed statistics of these datasets. Since the size of all English Wikipedia article is too large for some baselines, either on time or memory, we randomly sample 10 million documents as the *English* dataset; the *Spanish* dataset has more than 11 million documents and about 1 billion tokens; the smallest dataset, *Chinese*, is still more than 1.5 GB in file size.

Table 3.1: Dataset statistics

Language	# Docs	# Tokens	File Size
English	10,000,000	808,013,246	3.94 GB
Spanish	11,032,323	791,230,831	4.06 GB
Chinese	241,291	371,899,237	1.56 GB

We compare AutoPhrase+ with three types of methods as follows. Every method returns a ranked list of phrases.

Phrase Mining: There are many phrase extraction methods, such as NLP chunking methods, ConExtr [Parameswaran et al., 2010], KEA [Witten et al., 1999], TopMine [El-Kishky et al., 2015], and SegPhrase+. SegPhrase+ has shown its advantage over others.

- **SegPhrase+**: SegPhrase+ is for *English* corpus, which outperformed many other phrase mining, keyphrase extraction, and noun phrase chunking methods. However, it requires the effort of human experts to label hundreds of binary phrase. To adapt this work to the automated phrase mining setting in this paper, we feed the binary phrase labels used by AutoPhrase+ to SegPhrase+.

- **WrapSegPhrae**[2]: Moreover, to make SegPhrase+ support different languages, we add an encoding preprocessing to first transform non-English corpus using English characters and punctuation as well as a decoding postprocessing to translate them to the original language.

Parser-based Phrase Extraction: Using language-specific parsers, we can extract the minimum phrase units (e.g., NP) from the parsing trees as phrase candidates. Parsers of all three languages are available in Stanford NLP tools [De Marneffe et al., 2006, Levy and Manning, 2003, Nivre et al., 2016]. Two ranking heuristics are considered.

- **TF-IDF**: Rank the extracted phrase candidates by TF-IDF. It is more effective than C-Value as shown in Liu et al. [2015].

- **TextRank**: An unsupervised graph-based ranking model for keyword extraction [Mihalcea and Tarau, 2004].

Chinese Segmentation Models: Different from English and Spanish, phrasal segmentation in Chinese has been intensively studied because there is no whitespace in Chinese. The most effective and popular segmentation methods are the following.

- **AnsjSeg**[3] is a popular text segmentation algorithm for Chinese corpus. It ensembles statistical modeling methods of Conditional Random Fields (CRF) and Hidden Markov Models (HMMs) based on the *n*-gram setting.

- **JiebaPSeg**[4] is a *Chinese* text segmentation method implemented in Python. It builds a directed acyclic graph for all possible phrase combinations based on a prefix dictionary structure to achieve efficient phrase graph scanning. Then it uses dynamic programming to find the most probable combination based on the phrase frequency. For unknown phrases, an HMM-based model is used with the Viterbi algorithm.

---

[2]https://github.com/remenberl/SegPhrase-MultiLingual
[3]https://github.com/NLPchina/ansj_seg
[4]https://github.com/fxsjy/jieba

Note that all parser-based phrase extraction and Chinese segmentation models are pre-trained based on general corpus, which should be similar to our Wikipedia datasets.

We denote our proposed method as **AutoPhrase+**. If only the sub-ranked list of multi-word phrases in AutoPhrase+ is returned, it degenerates to **AutoPhrase**. If the context-aware phrasal segmentation degenerates to the length penalty mode (i.e., all $\delta(t_1, t_2)$ share the same value), we name it as **AutoSegPhrase**.

3.3.1   EXPERIMENTAL SETTINGS

Default Parameters.   We set the minimum support threshold σ as 30 and the maximum phrase length as 6, which are two parameters required by all methods. Other parameters required by compared methods were set according to the open-source tools or the original papers.

Efficiency Testing Environment.   The following execution time experiments were all conducted on the same machine mentioned in the previous chapter. The algorithm is fully implemented in C++. The preprocessing includes tokenizers from Lucene and Stanford NLP as well as the POS tagger from TreeTagger.

3.3.2   QUANTITATIVE EVALUATION AND RESULTS

Our experiments are designed to study how well our methods perform in terms of *precision* and *recall* compared to other methods. For a list of phrases, precision is defined as the number of occurred quality phrases divided by the length of the list. Recall is defined as the number of occurred quality phrases divided by the total number of quality phrases. For a ranked list, when a quality phrase encountered, we record the precision and recall of the prefix ranked list. In the end, we evaluate the precision-recall curves using the records.

Human Annotation. We rely on human evaluators to judge the quality of the phrases which cannot be identified through any knowledge base. More specifically, on each dataset, we randomly sample 500 such phrases from the predicted phrases of each method in the experiments. These selected phrases are shuffled in a shared *pool* and evaluated by three reviewers independently. We allow reviewers to use search engines when unfamiliar phrases encountered. By the rule of majority voting, phrases in this pool received at least two positive annotations are *quality phrases*. The intra-class correlations (ICCs) are all more than 0.9 on all datasets, which shows the agreement. focus on evaluating the ranked list in the pool.

Precision-recall curves of all compared methods evaluated by human annotation on three datasets are presented in Figure 3.5. The trends on the *English* and *Spanish* datasets are similar, while the trend on the *Chinese* dataset is slightly different.

AutoPhrase+ is the best among all compared methods on all datasets in terms of not only precision but also recall. Significant recall advantages can be always observed on all *English*, *Spanish*, and *Chinese* datasets regardless of using either Wiki phrases or human annotation for evaluation. For example, on the *English* dataset, the recall of AutoPhrase+ is more than 20% higher than the

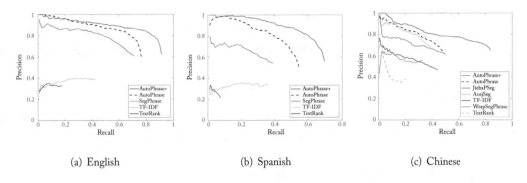

(a) English   (b) Spanish   (c) Chinese

Figure 3.5: Precision-recall curves evaluated by human annotation.

second best method (SegPhrase+) in absolute value when evaluating by Wiki phrases. Moreover, the recall differences between AutoPhrase+ and its variant AutoPhrase, ranging from 10% to 30% sheds light on the importance of modeling single-word phrases. Meanwhile, one can also observe that there is always a big precision gap between AutoPhrase+ and the best baseline on all three datasets. Without any surprise, the phrase chunking-based methods TF-IDF and TextRank work poorly, because the extraction and ranking are separated instead of unified.

Across two Latin language datasets, *English* and *Spanish*, the precision-recall curves of different methods are in the similar shapes. AutoPhrase+ and AutoPhrase overlaps in the beginning, but later, the precision of AutoPhrase drops earlier and has a lower recall due to the lack of single-word phrases. However, AutoPhrase works better than the previous state-of-the-art method SegPhrase+. TextRank starts with a higher precision than TF-IDF, but its recall is very low because of the sparsity of the constructed co-occurrence graph. TF-IDF achieves a reasonable recall but unsatisfactory precision.

On *Chinese* dataset, AutoPhrase+ and AutoPhrase has a clear gap even in the very beginning, which is different from the trends on the *English* and *Spanish* datasets, which reflects that single-word phrases are more important in Chinese. The major reason behind is that there are a considerable number of high-quality phrases (e.g., person names) in Chinese have only one token after tokenization. The performance of Chinese segmentation model AnsjSeg is very competitive, which is slightly better than WrapSegPhrase especially when evaluating by human annotation and shows comparable performance as AutoPhrase. This is because it not only leverages training data for segmentations, but also exhausts the engineering work, including a huge dictionary for popular Chinese entity names and specific rules for certain types of entities. As a consequence, AnsjSeg can easily extract tons of well-known terms and people/location names. Outperforming such a strong baseline further confirms the effectiveness of AutoPhrase+. TF-IDF is slightly better than another pre-trained Chinese segmentation method JiebaPSeg, while TextRank works worst again.

In conclusion, our proposed AutoPhrase+ consistently works the best among all compared methods and thus demonstrating its effectiveness on three datasets in different languages. The difference between AutoPhrase+ and AutoPhrase shows the necessity of modeling single-word phrases.

3.3.3   DISTANT TRAINING EXPLORATION

To compare the distant training and domain expert labeling, we introduce two domain-specific datasets in English: *DBLP* and *Yelp* as shown in the following table.

Table 3.2: Two domain-specific datasets in English

Dataset	Domain	# of Tokens	File Size	Positive Pool Size
DBLP	Scientific Paper	91.6 M	618 MB	29 K
Yelp	Business Review	145.1 M	749 MB	22 K

To be fair, all the configurations in the classifiers are the same except for the label selection process. More specifically, we come up with four training pools:

1. **EP** means that domain experts give the positive pool.
2. **DP** means that a sampled subset from existing general knowledge forms the positive pool.
3. **EN** means that domain experts give the negative pool.
4. **DN** means that all *unlabeled* (i.e., not in the positive pool) phrase candidates form the negative pool.

By combining any pair of the positive and negative pools, we have four variants, **EPEN** (in SegPhrase+), **DPDN** (in AutoPhrase+), **EPDN**, and **DPEN**.

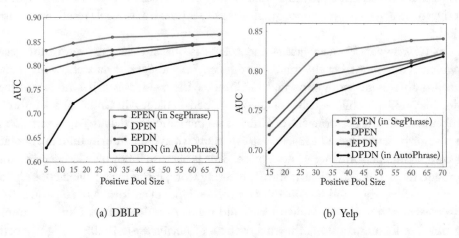

(a) DBLP   (b) Yelp

Figure 3.6: AUC curves of four variants *when we have enough positive labels in the positive pool* **EP**.

First of all, we evaluate the performance difference in the two positive pools. Compared to EPEN, DPEN adopts a positive pool sampled from knowledge bases instead of the well-designed positive pool given by domain experts. The negative pool *EN* is shared. As shown in Figure 3.6, we vary the size of the positive pool and plot their AUC curves. We can find that EPEN outperforms DPEN and the trends of curves on both datasets are similar. Therefore, we conclude that the positive pool generated from knowledge bases has reasonable quality, although its corresponding quality estimator works slightly worse.

Secondly, we verify that whether the proposed noise reduction mechanism works properly. Compared to EPEN, EPDN adopts a negative pool of all unlabeled phrase candidates instead of the well-designed negative pool given by domain experts. The positive pool *EP* is shared. In Figure 3.6, the clear gap between them and the similar trends on both datasets show that the noisy negative pool is slightly worse than the well-designed negative pool, but it still works effectively.

As illustrated in Figure 3.6, DPDN has the worst performance when the size of positive pools are limited. However, distant training can generate much larger positive pools, which may significantly beyond the ability of domain experts considering the high expense of labeling. Consequently, we are curious whether the distant training can finally beat domain experts when positive pool sizes become large enough. We call the size at this tipping point as the ideal number.

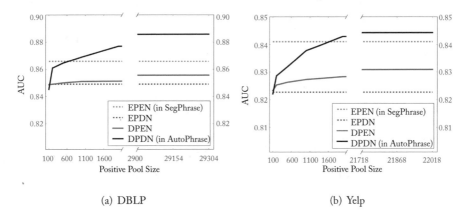

(a) DBLP   (b) Yelp

Figure 3.7: AUC curves of four variants *after we exhaust positive labels in the positive pool* **EP**.

We increase positive pool sizes and plot AUC curves of DPEN and DPDN, while EPEN and EPDN are degenerated as dashed lines due to the limited domain expert abilities. As shown in Figure 3.7, with a large enough positive pool, distant training is able to beat expert labeling. On the *DBLP* dataset, the ideal number is about 700, while on the *Yelp* dataset, it becomes around 1600. Our guess is that the ideal training size is proportional to the number of words (e.g., 91.6M in *DBLP* and 145.1M in *Yelp*). We notice that compared to the corpus size, the ideal number is relatively small, which implies the distant training should be effective in many domain-specific corpora as if they overlap with Wikipedia.

Besides, Figure 3.7 shows that when the positive pool size continues growing, the AUC score increases but the slope becomes smaller. The performance of distant training will be finally stable when a relatively large number of quality phrases were fed.

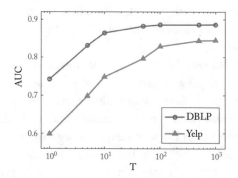

Figure 3.8: AUC curves of DPDN varying T.

We are curious how many trees (i.e., T) is enough for DPDN. We increase T and plot AUC curves of DPDN. As shown in Figure 3.8, on both datasets, as T grows, the AUC scores first increase rapidly and later the speed slows down gradually, which is consistent with the theoretical analysis in Section 3.2.1.

3.3.4   POS-GUIDED PHRASAL SEGMENTATION

We are also interested in how much performance gain we can obtain from incorporating POS tags in this segmentation model, especially for different languages. We select Wikipedia article datasets in three different languages: *English*, *Spanish*, and *Chinese*. To be fair, since SegPhrase+ only models multi-word phrases, we only use AutoPhrase for the comparison.

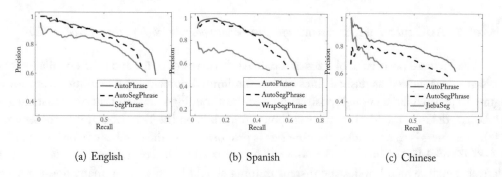

(a) English   (b) Spanish   (c) Chinese

Figure 3.9: Precision-recall curves of AutoPhrase and AutoSegPhrase.

Figure 3.9 compares the results of AutoPhrase and AutoSegPhrase, with the best baseline methods as references. AutoPhrase outperforms AutoSegPhrase even on the *English* dataset, though it has been shown in the last chapter the length penalty works reasonably well for English. The *Spanish* dataset has similar observation. Moreover, the advantage of AutoPhrase becomes more significant on the *Chinese* dataset, indicating the poor generality of length penalty.

In summary, thanks to the extra context information and syntactic information for the particular language, incorporating POS tags during the phrasal segmentation can work better than equally penalizing phrases of the same length.

3.3.5   EFFICIENCY STUDY

Figures 3.10a and 3.10b evaluate the running time and the peak memory usage of AutoPhrase+ using 10 threads on different proportions of three datasets, respectively. Both time and memory are linear to the size of text corpora. Moreover, AutoPhrase+ can also be parallelized in an almost lock-free way and shows a linear speedup in Figure 3.10c.

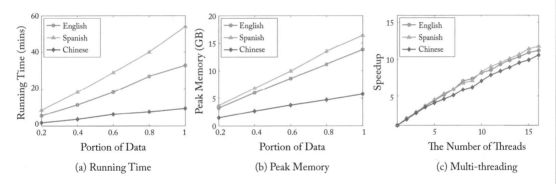

Figure 3.10: Efficiency of AutoPhrase+.

Besides, compared to the previous state-of-the-art phrase mining method SegPhrase+ and its variants WrapSegPhrase on three datasets, as shown in Table 3.3, AutoPhrase+ achieves about 8 to 11 times speedup and about 5 to 7 times memory usage improvement. These improvements are made by a more efficient indexing and a more thorough parallelization.

3.3.6   CASE STUDY

We present a case study about the extracted phrases as shown in Table 3.4. The top ranked phrases are mostly named entities, which makes sense for the Wikipedia article datasets. Even in the long tail part, there are still many high-quality phrases. For example, we have the 「great spotted woodpecker」 (a type of birds) and 「计算机 科学技术」 (i.e., Computer Science and Technology) ranked about 100,000. In fact, we have more than 345K and 116K phrases with a phrase quality higher than 0.5 on the *EN* and *CN* datasets, respectively.

Table 3.3: Efficiency comparison between AutoPhrase+ and SegPhrase+/WrapSegPhrase utilizing 10 threads

	English		Spanish		Chinese	
	Time (mins)	Memory (GB)	Time (mins)	Memory (GB)	Time (mins)	Memory (GB)
AutoPhrase+	32.77	13.77	54.05	16.42	9.43	5.75
(Wrap)SegPhrase	369.53	97.72	452.85	92.47	108.58	35.38
Speed/Saving	11.27	86%	8.37	82%	11.50	83%

Table 3.4: The results of AutoPhrase+ on the *EN* and *CN* datasets, with translations and explanations for Chinese phrases. The whitespaces on the *CN* dataset are inserted by the Chinese tokenizer.

	EN	CN	
Rank	Phrase	Phrase	Translation (Explanation)
1	Elf Aquitaine	江苏舜天	(the name of a soccer team)
2	Arnold Sommerfeld	苦艾酒	Absinthe
3	Eugene Wigner	白发魔女	(the name of a novel or a TV-series)
4	Tarpon Springs	笔记型电脑	notebook computer, laptop
5	Sean Astin	党委书记	Secretary of Party Committee
...	...	...	...
20,001	ECAC Hockey	非洲国家	Aftican countries
20,002	Sacramento Bee	左翼党	The Left (German: Die Linke)
20,003	Bering Strait	菲沙河谷	Fraser Valley
20,004	Jacknife Lee	海马体	Hippocampus
20,005	WXYZ-TV	斋贺光希	Mitsuki Saiga (a voice actress)
...	...	...	...
99,994	John Gregson	计算机科学技术	Computer Science and Technology
99,995	white-tailed eagle	恒天然	Fonterra (a company)
99,996	rhombic dodecahedron	中国作家协会 副主席	The Vice President of Writers Association of China
99,997	great spotted woodpecker	维他命b	Vitamin B
99,998	David Manners	舆论导向	controlled guidance of the media
...			

CHAPTER 4

Phrase Mining Applications

This book investigates the problem of phrase mining and introduces a series of methodologies to solve it. It first presents the limitation of relying on n-gram-based representation, and then proposes to use quality phrases as the representation units due to its superior intepretability.

Both effective and scalable solutions are proposed and empirically validated on multiple real world datasets, such as scientific publications, business reviews, and online encyclopedia. The corresponding source code has been released on Github and we encourage open collaborative development.

- **SegPhrase+** Liu et al. [2015]: https://github.com/shangjingbo1226/SegPhrase

- **AutoPhrase+** Shang et al. [2017]: https://github.com/shangjingbo1226/AutoPhrase

In the following sections, we introduce four applications to showcase the impact of the phrase mining results, and discuss the research frontier.

4.1 LATENT KEYPHRASE INFERENCE

Quality phrases mined in previous chapters are document-independent. That is to say, regarding a particular document, it is difficult to tell which phrase is more salient than the rest. One solution studied in our recent publication [Liu et al., 2016] is to rank phrases by their topical relevance with the document content. In particular, this work is motivated by the application of document representation.

If one looks back in the literature, the most common document representation is the bag-of-words due to its simplicity and efficiency. This method, however, typically fails to capture word-level synonymy (missing shared concepts in distinct words, such as "doctor" and "physician") and polysemy (missing distinct concepts in same word, such as "Washington" can be either the city or the government). As a remedy, topic models [Blei et al., 2003, Deerwester et al., 1990] try to overcome this limitation by positing a set of latent topics which are distributions over words, and assuming that each document can be described as a mixture of these topics. Nevertheless, the interpretability of latent space for topic models is not straightforward and pursuing semantic meaning in inferred topics is difficult. Concept-based models [Gabrilovich and Markovitch, 2007, Gottron et al., 2011, Hassan and Mihalcea, 2011, Song et al., 2011] were proposed to overcome these barriers. The intuition is to link the documents with concepts in a general Knowledge Base (KB), like Wikipedia or Freebase, and assign relevance score accordingly.

For example, the text sequence "*DBSCAN for knowledge discovery*" can be mapped to KB concepts like "*KB: data mining,*" "*KB: density-based clustering*" and "*KB: dbscan*" (relevance scores are omitted). Such methods take advantage of a vast amount of highly organized human knowledge. However, most of the existing knowledge bases are manually maintained, and are limited in coverage and freshness. Researchers have therefore recently developed systems such as Probase [Wu et al., 2012] and DBpedia [Bizer et al., 2009] to replace or enrich traditional KBs. Neverthless, the rapid emergence of large, domain-specific text corpora (e.g., business reviews) poses significant challenges to traditional concept-based techniques and calls for methods of representing documents by interpretable units without requirement of a KB.

One solution in Liu et al. [2016] is to instantiate the interpretable units in the document representation as *quality phrases*. That is to say, a document is represented as a subset of quality phrases that are informative to summarize the document content. For ease of presentation, we name these phrases as *document keyphrases*.

However, not all document keyphrases are frequently mentioned in the text. In other words, phrase frequency does not necessarily indicate the saliency. To deal with this challenge, we propose to associate each quality phrase with a *silhouette*—a cohesive set of topically related content units (i.e., words and phrases), which is learned from the corpus itself, to help infer topical relevance between the document content and each quality phrase mined in previous chapters.

These silhouettes also enhance the interpretability of corresponding quality phrases. An example of using document keyphrases to represent text is provided in Table 4.1, together with results of other approaches. Its underlying technique, called *Latent Keyphrase Inference* (LAKI) is shown in Figure 4.1. LAKI can be divided into two phases: (i) *the offline phrase silhouette learning phase*, which extracts quality phrases from the in-domain corpus and learns their silhouettes respectively, and (ii) *the online document keyphrase inference phase*, which identify keyphrases for each query based on the quality phrase silhouettes, as outlined below.

Table 4.1: Representations for query "DBSCAN is a method for clustering in process of knowledge discovery," returned by various categories of methods

Categories	Representation
Words	dbscan, method, clustering, process, ...
Topics	[k-means, clustering, clusters, dbscan, ...]
	[clusters, density, dbscan, clustering, ...]
	[machine, learning, knowledge, mining, ...]
KB Concepts	data mining, clustering analysis, dbscan, ...
Document Keyphrases	dbscan: [dbscan, density, clustering, ...]
	clustering: [clustering, clusters, partition, ...]
	data mining: [data mining, knowledge, ...]

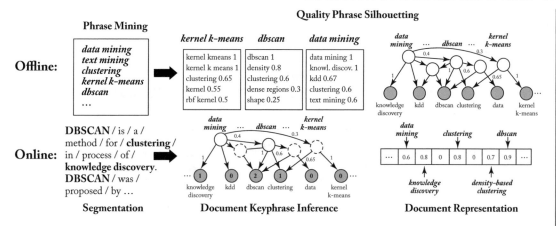

Figure 4.1: Overview of LAKI. White and grey nodes represent quality phrases and content units, respectively.

- *Offline Phrase Silhouette Learning:*
    1. Mine quality phrases from a textual corpus; and
    2. learn quality phrase silhouettes by iteratively optimizing a Bayesian network with respect to the unknown values, i.e., latent document keyphrases, given observed content units in the training corpus.

- *Online Document Keyphrase Inference:*
    1. Segment input query into content units; and
    2. do inference for document keyphrases given the observed content units, which quantifies relatedness between the input query and corresponding keyphrase.

The offline phase is critical in the sense that the online inference phase can be formulated as its sub-process. Technically speaking, the learning is done by optimizing a statistical Bayesian network, given observed content units (i.e., words and phrases after phrasal segmentation) in a training corpus. We use a DAG-like Bayesian network shown in Figure 4.2. Content units are located at the bottom layer and quality phrases form the rest. Both types of nodes act as binary variables[1] and directional links between nodes depict their dependency.

Before diving into the details, we motivate our Bayesian approach to the silhouetting problem. First, this approach enables our model to infer not just explicitly mentioned document keyphrases. For example, even if the text only contains "html" and "css," the words "web page" come to mind. But more than that, a multi-layered network will activate an ancestor-quality

---

[1] For multiple mentions of a content unit, we can simply make several copies of that node together with its links.

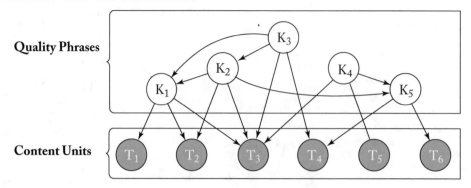

Figure 4.2: An illustrative Bayesian network for quality phrase silhouetting.

phrase like "world wide web" even they are not directly linked to "html" or "css," which are content units in the bottom layer.

Meanwhile, we expect to identify document keyphrases with different relatedness scores. Reflected in this Bayesian model from a top-down view, when a parent quality phrase is activated, it is more possible for its children with stronger connection to get activated.

Furthermore, this formulation is flexible. We allow a content unit to get activated by each connected quality phrase as well as by a random noise factor (not shown in Figure 4.2), behaving like a *Noisy-OR*, i.e., a logical OR gate with some probability of having "noisy" output. This increases robustness of the model especially when training documents are noisy.

There are two challenges in the learning process: (1) how to learn link weights given the fixed Bayesian network structure and (2) how the initialization is done to decide this structure and to set initial link weights.

For the former, to effectively learn link weights, Maximum Likelihood Estimation (MLE) is adopted. The intuition is to maximize the likelihood of observing content units together with partially-observed document keyphrases.[2] But it is extremely difficult to directly optimize due to the latent states for the rest quality phrases. In this case, we usually resort to the Expectation-Maximization (EM) algorithm which guarantees to give a local optimum solution. The EM algorithm starts with some initial guess at the link weights and then proceeds to iteratively generate successive estimates by repeatedly applying the E-step (Expectation-step) and M-step (Maximization-step) until the MLE objective changes minimally.

Expectation Step: The whole E-step is trying to compute the conditional probability of unobserved document keyphrases considering all their state combinations. It turns out that this step is exactly the same as what we conduct in the online inference phase. In other words, the online inference phase is just a sub-process of the offline training phase.

---

[2]Explicit document keyphrases can be identified by applying existing keyphrase extraction methods like Witten et al. [1999].

Unfortunately, the E-step cannot be easily executed. Since each latent quality phrase in Figure 4.2 acts as a binary variable, the size of possible state combinations can be as big as $O(2^n)$. That is to say, to accurately compute the probabilities required in E-step is NP-hard for a Bayesian network like ours [Cooper, 1990]. We therefore adopt two approaches to approximately collect sufficient statistics. The first idea is to apply sampling technique such as Markov Chain Monte Carlo to search for the most likely state combinations. Among the Monte Carlo family, we apply Gibbs sampling in this work to sample quality phrase variables during each E-step. Given content unit vector representing a document, we proceed as follows.

1. Start with initial setting: only observed content units and explicit document keyphrases are set to be true.

2. For each sampling step, sequentially sample each quality phrase node following conditional distribution of that node given all other nodes with fixed states.

The above Gibbs sampling process ensures that samples approximate the joint probability distribution between all phrase variables and content units.

The second approach is applied as the preprocessing right before the E-step. The idea is to exclude non-related quality phrases that we are confident with. Intuitively, only a small portion of quality phrases are related to the observed text. There is no need to sample all phrase nodes since most of them do not have chance to get activated. That is to say, we can skip majority of them based on a reasonable relatedness prediction before conducting Gibbs sampling. We adopt a local arborescence structure [Wang et al., 2012] to approximate the original Bayesian network which allows us to roughly approximate the score for each node in an efficient way. We opt to omit the technical details here and interesting readers can refer to the original paper [Liu et al., 2016].

Maximization Step: The M-step tries to update link weight based on the sufficient statistics collected by the Exspectation step. In this problem setting, we are able to obtain a closed form solution by taking the derivative of the MLE objective function.

Now the rest challenge is to decide the Bayesian network structure and to set initial link weights. A reasonable topological order of DAG should be similar to that of a domain ontology. The links among quality phrase nodes should reflect IS-A relationships [Yin and Shah, 2010]. Ideally, documents which are describing specific topics will first imply some deep quality phrase nodes being activated. Then the ontology-like topological order ensures these content units have the chance of being jointly activated by general phrase nodes via inter-phrase links. Many techniques [Dahab et al., 2008, Sanderson and Croft, 1999, Yin and Shah, 2010] have been previously developed to induce an ontological structure over quality phrases. It is out of scope of our work to specifically address these or evaluate their relative impact in our evaluation. We instead use a simple data-driven approach, where quality phrases are sorted based on their counts in the corpus, assuming phrase generality is positively correlated with its number of mentions. Thus, quality phrases mentioned more often are higher up in the graph. Links are added between quality phrases when they are closely related and frequently co-occurred. The link weights between nodes are simply set to be their similarity scores computed from the Word2Vec [Mikolov et al., 2013].

To verify the effectiveness of LAKI, we present several queries with their top-ranked document keyphrases in Table 4.2 generated from the online phase of LAKI. Overall, we see that the method can handle both short and long queries quite well. Most document keyphrases are successfully identified in the list. Relatedness between keyphrase and queries generally drops with ranking lowers down. Meanwhile, both general and specific document keyphrases exist in the ranked list. This provides LAKI with more discriminative power when someone applies it to text mining applications like document clustering and classification. Moreover, the method has the ability to process ambiguous queries like "lda" based on contextual words "topic." We attribute this to the well-modeled quality phrase silhouettes and we show some examples of them in Table 4.3. As a quality phrase silhouette might contain many content units, we only demonstrate ones with the most significant link weights. For ease of presentation, link weights are omitted in the table.

4.2 TOPIC EXPLORATION FOR DOCUMENT COLLECTION

The previous application targets at single document analysis. How do we deal with a collection of documents?

Most textbooks might tell you to build a topic model like Latent Dirichlet Allocation [Blei et al., 2003]. It assumes that each document can be modeled as a mixture of latent topics and each topic is represented as a mixture of unigrams. One can certainly replace the unigram-based text input [El-Kishky et al., 2015, Guan et al., 2016, Wang et al., 2013] with our phrasal segmentation results such that topics become a mixture of phrases.

A more interesting approach for topic exploration is inspired from word embedding [Mikolov et al., 2013]. We have shown some case studies in Section 2.4.4 that, after applying the algorithm, words and phrases are mapped into a vector space such that semantically similar words and phrases have similar vector representations. On top of that, one can apply hierarchical clustering over these embeddings, with the intuition that depth of the clusters in the hierarchy implies topic granularity.

Recently, TripAdvisor has published a tech article[3] at its blog site introducing their application built on hotel reviews. With the help of phrase mining and word embedding, they are able to apply agglomerative clustering to obtain a gigantic hierarchy. Figure 4.3 shows a toy hierarchy clustered over 18 phrases. Up to this point the process is fully automatic. The remaining step is to rely on human curation to pick out the interesting clusters, and give each of them a snappy name. The manual process is fairly easy. Given a list of phrases for a cluster, together with hotels that are the most related to the cluster, and even some examples sentences from the review text for

---

[3]http://engineering.tripadvisor.com/using-nlp-to-find-interesting-collections-of-hotels/

Table 4.2: Examples of document representation by LAKI with top-ranked document keyphrases (relatedness scores are ommited due to the space limit)

Query	LDA	BOA
Document Keyphrases	linear discriminant analysis, latent dirichlet allocation, topic models, topic modeling, face recognition, latent dirichlet, generative model, topic, subspace models, . . .	boa steakhouse, bank of america, stripsteak, agnolotti, credit card, santa monica, restaurants, wells fargo, steakhouse, prime rib, bank, vegas, las vegas, cash, cut, dinner, bank, money, . . .
Query	**LDA topic**	**BOA steak**
Document Keyphrases	latent dirichlet allocation, topic, topic models, topic modeling, probabilistic topic models, latent topics, topic discovery, generative model, mixture, text mining, topic distribution, etc.	steak, stripsteak, boa steakhouse, steakhouse, ribeye, craftsteak, santa monica, medium rare, prime, vegas, entrees, potatoes, french fries, filet mignon, mashed potatoes, texas roadhouse, etc.
Query	**SVM**	**deep dish pizza**
Document Keyphrases	support vector machines, svm classifier, multi class, training set, margin, knn, classification problems, kernel function, multi class svm, multi class support vector machine, support vector, etc.	deep dish pizza, chicago, deep dish, amore taste of chicago, amore, pizza, oregano, chicago style, chicago style deep dish pizza, thin crust, windy city, slice, pan, oven, pepperoni, hot dog, etc.
Query	**Mining Frequent Patterns without Candidate Generation**	**I am a huge fan of the All You Can Eat Chinese food buffet**
Document Keyphrases	mining frequent patterns, candidate generation, frequent pattern mining, candidate, prune, fp growth, frequent pattern tree, apriori, subtrees, frequent patterns, candidate sets, etc.	all you can eat, chinese food, buffet, chinese buffet, dim sum, orange chicken, chinese restaurant, asian food, asian buffet, crab legs, lunch buffet, fan, salad bar, all you can drink, etc.
Query	**Text mining, also referred to as text data mining, roughly equivalent to text analytics, refers to the process of deriving high-quality information from text. High-quality information is typically derived through means such as statistical pattern learning.**	**It's the perfect steakhouse for both meat and fish lovers. My table guest was completely delirious about his Kobe Beef and my lobster was perfectly cooked. Good wine list, they have a lovely Sancerre! Professional staff, quick and smooth.**
Document Keyphrases	text analytics, text mining, patterns, text, textual data, topic, information, text documents, information extraction, machine learning, data mining, knowledge discovery, etc.	kobe beef, fish lovers, steakhouse, sancerre, wine list, guests, perfectly cooked, lobster, staff, meat, fillet, fish, lover, seafood, ribeye, filet, sea bass, risotto, starter, scallops, steak, beef, etc.
	Academia	Yelp

Table 4.3: Examples of quality phrase silhouettes (from offline quality phrase silhouette learning). Link weights are omitted.

Quality Phrase	linear discriminant analysis	boa steakhouse
Silhouette	linear discriminant analysis, lda, face recognition, feature extraction, principle component analysis, uncorrelated, between class scatter, etc.	boa steakhouse, boa, steakhouse, restaurant, dinner, strip steak, craftsteak, santa monica, vegas, filet, ribeye, new york strip, sushi roku, etc.
Quality Phrase	latent dirichlet allocation	ribeye
Silhouette	latent dirichlet allocation, lda, topics, perplexity, variants, subspace, mixture, baselines, topic models, text mining, bag of words, etc.	ribeye, steak, medium rare, medium, oz, marbled, new york strip, well done, prime rib, fatty, juicy, top sirloin, filet mignon, fillet, etc.
Quality Phrase	support vector machines	deep dish
Silhouette	support vector machines, svm, classification, training, classifier, machine learning, prediction, hybrid, kernel, feature selection, etc.	deep dish, pizza, crust, thin crust pizza, chicago, slice, pepperoni, deep dish pizza, pan style, pizza joints, oregano, stuffed crust, chicago style, etc.
Quality Phrase	fp growth	chinese food
Silhouette	fp growth, algorithm, apriori like, mining, apriori, frequent patterns, mining association rules, frequent pattern mining, fp tree, etc.	chinese food, food, chinese, restaurants, americanized, asian, orange chicken, chow mein, wok, dim sum, panda express, chinese cuisine, etc.
Quality Phrase	text mining	mcdonalds
Silhouette	text mining, text, information retrieval, machine learning, topics, knowledge discovery, text data mining, text clustering, nlp, etc.	mcdonalds, drive through, fast food, mcnugget, mcflurry, fast food chain, sausage mcmuffin, big bag, mcmuffin, burger king, etc.
Quality Phrase	database	sushi
Silhouette	database, information, set, objects, storing, retrieval, queries, accessing, relational, indexing, record, tables, query processing, transactions, etc.	sushi, rolls, japanese, sushi joint, seafood, ayce, sushi rolls, salmon sushi, tuna sushi, california roll, sashimi, sushi lovers, sushi fish, etc.
	Academia	Yelp

those hotels, one can determine if it would a good way to explore the hotels of a city. It would be difficult to mathematically define what is interesting, but easy for a human to know when they see it. The human can also come up with a clever name, which is also simple given the list of quality phrases.

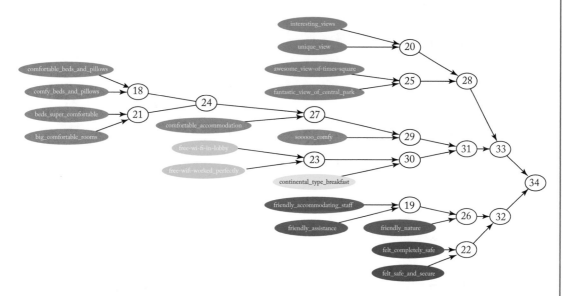

Figure 4.3: Agglomerative clustering over 18 phrases.

Some interesting collections are shown in Figure 4.4. The whole process provides insight into a particular city, picking out interesting neighborhoods, features of the hotels, and nearby attractions.

To systematically analyze large numbers of textual documents, another approach is to manage documents (and their associated metadata) in a multi-dimensional fashion (e.g., document category, date/time, location, author, etc.). Such structure provides flexibility of understanding local information with different granularities. Moreover, the contextualized analysis often yields comparative insights. That is, given a pair of document collections, how to identify common and different content units of various granularity (e.g., words, sentences).

However, word-based summarization suffers from limited readability as single words are usually non-informative and bag-of-words representation does not capture the semantics of the original document well—it may not be easy for users to interpret the combined meaning of the words. Sentence-based summarization, on the other hand, may be too verbose to highlight the general commonalities and differences—users may be distracted by the irrelevant information contained there. Recent studies [Ren et al., 2017a, Tao et al., 2016] leverage quality phrases, i.e., minimal semantic unit, to summarize the commonalities and differences. Figure 4.5 gives an example where an analyst may pose multidimensional queries and the system is able to leverage

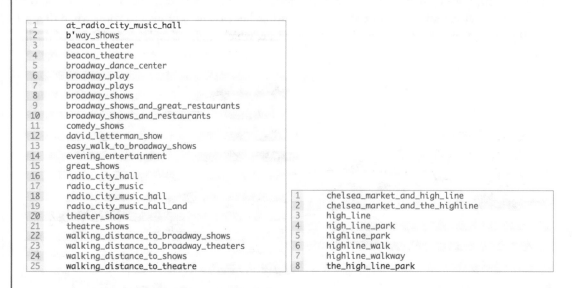

1	at_radio_city_music_hall
2	b'way_shows
3	beacon_theater
4	beacon_theatre
5	broadway_dance_center
6	broadway_play
7	broadway_plays
8	broadway_shows
9	broadway_shows_and_great_restaurants
10	broadway_shows_and_restaurants
11	comedy_shows
12	david_letterman_show
13	easy_walk_to_broadway_shows
14	evening_entertainment
15	great_shows
16	radio_city_hall
17	radio_city_music
18	radio_city_music_hall
19	radio_city_music_hall_and
20	theater_shows
21	theatre_shows
22	walking_distance_to_broadway_shows
23	walking_distance_to_broadway_theaters
24	walking_distance_to_shows
25	walking_distance_to_theatre

1	chelsea_market_and_high_line
2	chelsea_market_and_the_highline
3	high_line
4	high_line_park
5	highline_park
6	highline_walk
7	highline_walkway
8	the_high_line_park

Figure 4.4: Left: collection "catch a show;" Right: collection "near the high line."

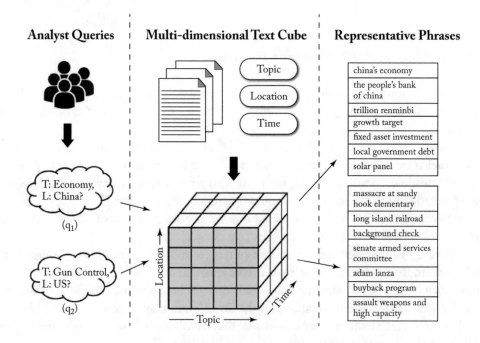

Figure 4.5: Illustration of phrase-based summarization.

the relation between document subsets induced by query context and identify phrases that truly distinguish the queried subset of documents from neighboring subsets.

Example 4.1   Suppose a multi-dimensional text database is constructed from *The New York Times* news repository with three meta attributes: *Location*, *Topic*, and *Time*, as shown in Figure 4.5. An analyst may pose multidimensional queries such as: (q1): ⟨China, Economy⟩ and (q2): ⟨US, Gun Control⟩. Each query asks for summary of a cell defined by two dimensions Location and Topic. What kind of cell summary does she like to see? Frequent unigrams such as debt or senate are not as informative as multi-word phrases, such as local government debt and senate armed service committee. The phrases preserve better semantics as integral units rather than as separate words.

Generally, three criteria should be considered when ranking representative phrases in a selected multidimensional cell: (i) integrity: a phrase that provides integral semantic unit should be preferable over nonintegral unigrams; (ii) popularity: popular in the selected cell (i.e., selected subset of documents); and (iii) distinctiveness: distinguish the selected cell from other cells.

Within the whole ranked phrase list, top-k representative phrases normally have higher value for users in text analysis. As a further matter, the top-k query also enjoys computational superiority, so that users can conduct fast analysis.

Bearing all these in mind, the authors have designed statistical measures for each criterion, and uses geometric mean of those three scores as the ranking signal. The specific design principles are as follows.

1. Popularity and distinctiveness of a phrase are dependent of the target cell, while integrity is not.

2. Popularity and distinctiveness can be measured from frequency statistics of a phrase in each cell, while integrity cannot. To measure integrity, one needs to investigate each occurrence of the phrase and other phrases to determine whether that phrase is indeed an integral semantic unit. The quality score provided by SegPhrase+ and AutoPhrase+ is a good indicator.

3. Popularity relies on statistics from documents only within the cell, while distinctiveness relies on documents both in and out of the cell. The documents involved for distinctiveness measure calculation is defined as contrastive document set. In the particular algorithm design in the paper, sibling set of the query cell is used as contrastive document set.

The algorithm is applied on *The New York Times* 2013–2016 dataset and *PubMed*[4] Cardiac data with their representative phrase list in Table 4.4 and Table 4.5. In the book, it is reported that using phrases achieves the best trade-off between processing efficiency, storage cost, and summarization interpretability.

---

[4]PubMed is a free full-text archive of biomedical and life sciences journal literature.

Table 4.4: Top-10 representative phrases for *The New York Times* queries

<U.S., Gun Control>	<U.S., Immigration>	<U.S., Domestic Politics>	<U.S., Law and Crime>	<U.S., Military>
gun laws	immigration debate	gun laws	district attorney	sexual assault in the military
the national rifle association	border security	insurance plans	shot and killed	military prosecutors
gun rights	guest worker program	background check	federal court	armed services committee
background check	immigration legislation	health coverage	life in prison	armed forces
gun owners	undocumented immigrants	tax increases	death row	defense secretary
assault weapons ban	overhaul of the nation's immigration laws	the national rifle association	grand jury	military personnel
mass shootings	legal status	assault weapons ban	department of justice	sexually assaulted
high capacity magazines	path to citizenship	immigration debate	child abuse	fort meade
gun legislation	immigration status	the federal exchange	plea deal	private manning
gun control advocates	immigration reform	medicaid program	second degree murder	pentagon officials

Table 4.5: Top representative phrases for five cardiac diseases

<Cerebrovascular Accident>	<Ischemic Heart Disease>	<Cardiomyopathy>	<Arrhythmia>	<Valve Dysfunction>
alpha-galactosidase a	Cholesteryl ester transfer protein	Interferon gamma	Methionine synthase	Mineralocorticoid receptor
brain neurotrophic factor	apolipoprotein a-I	interleukin-4	ryanodine receptor 2	tropomyosin alpha-1 chain
tissue-type activator	integrin alpha-iib	interleukin-17a	potassium v.g. h member 2	elastin
apolipoprotein e	adiponectin	titin	inward rectifier channel 2	beta-2-glycoprotein 1
neurogenic l.n.h.p. 3	p2y purinoceptor 12	tumor necrosis factor	beta-2-glycoprotein 1	myosin-binding protein c

Next, we introduce a topic exploration approach proposed from the perspective of graph mining. In Gui et al. [2016], a gigantic heterogeneous information network is constructed by incorporating metadata like authors, venues, and categories, as shown in Figure 4.6. The heterogeneity comes from the multiple types of nodes and links. This network is then modeled with an embedding approach, i.e., nodes in the network including the document phrases and metadata are all projected into a common low-dimensional space such that different types of nodes can be compared in a homogeneous fashion. The embedding algorithm is designed to preserve the semantic similarity among multi-typed nodes such that nodes that are semantically similar will be close in the space, with the distance measured by cosine similarity, for instance.

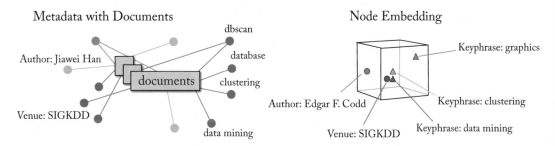

Figure 4.6: Embedding document keyphrases with metadata into the same space.

Different from the previous contrastive analysis, the embedding approach largely relies on the data redundancy to automatically infer the relatedness. By viewing the association between any document and its metadata as a link, the algorithm tries to push the embeddings of its constituent nodes closer to each other. For instance, observing the phrase "data mining" often appear together with venue "SIGKDD" rather than "SIGMOD" in the bibliographic corpus, the embedding distance between the pair "data mining" and "SIGKDD" should be smaller than that of "data mining" and "SIGMOD."

The underlying technique is essentially minimizing the Kullback-Leibler divergence between model distribution and empirical distribution defined on a target node out given the other participating nodes on the same link as context. Practically, one can solve the optimization problem by requiring the likelihood of an observed link to be higher than its corresponding "fake" link with one of the constituent node replaced by any other randomly sampled node. For more technical and experimental details, please refer to our recent paper [Gui et al., 2016].

4.3 KNOWLEDGE BASE CONSTRUCTION

The above applications are focusing on document-related summarization considering phrase mention as the representation unit. Beyond that, a big opportunity of utilizing phrase mining is to recognize real-world entities, such as people, products and organizations, from massive amount but interrelated unstructured data. By mining token spans of phrase mentions in documents,

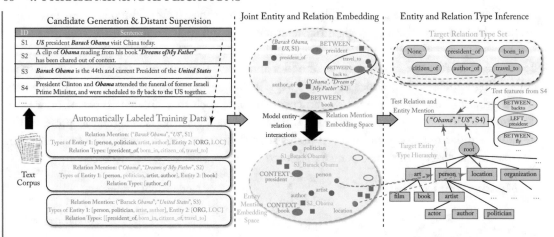

Figure 4.7: Framework overview of entity/relation joint typing.

labeling their structured types and inferring their relations, it is possible to construct or enrich semantically rich knowledge base and provide conceptual summarization of such data.

Existing entity detection tools such as noun phrase chunkers are trained on general-domain corpora (e.g., news articles), but they do not work effectively nor efficiently on domain-specific corpora such as Yelp reviews (e.g., "pulled pork sandwich" cannot be easily detected). Meanwhile, the process of manually labeling a training set with a large number of entity and relation types is too expensive and error-prone. Therefore, a major challenge is how to design domain-independent system that will apply to text corpora from different domains in the absence of human annotated, domain data. The rapid emergence of large, domain specific text corpora (e.g., news, scientific publications, social media content) calls for methods that can extract entities and relations of target types with minimal or no human supervision.

Recently, Ren et al. [2017b] introduced a novel entity/relation joint typing algorithm inspired from our corpus-scope phrase mining framework. The work extends our phrasal segmentation to detect entity mentions, and then jointly embeds entity mentions, relation mentions, text features, and type labels into two low-dimensional spaces (for entity and relation mentions respectively), where, in each space, objects whose types are close will also have similar representations.

Figure 4.7 illustrates the framework which comprises the following four major steps.

1. Run phrase mining algorithm on a textual corpus using positive examples obtained from an existing knowledge base, to detect candidate entity mentions.

2. Generate candidate relation mentions (sentences mentioning two candidate entities), extract text features for each relation mention and their entity mention argument. Apply distant supervision to generate labeled training data.

3. Jointly embed relation and entity mentions, text features, and type labels into two low-dimensional spaces (for entities and relations, respectively) where, in each space, close objects tend to share the same types.

4. Estimate type labels for each test relation mention and type-path for each test entity mention from learned embeddings, by performing nearest neighbor search in the target type set or the target type hierarchy.

Similar to AutoPhrase+, the first step utilizes POS tags and uses quality examples from KB as guidance to model the segment quality (i.e., "how likely a candidate segment is an entity mention"). But the detailed workflow is slightly different: (1) mine frequent contiguous patterns for both word sequence and POS tag sequence up to a fixed length from POS-tagged corpus; (2) extract features including corpus-level concordance and sentence-level lexical signals to train two random forest classifiers, for estimating quality of candidate phrase and candidate POS pattern; (3) find the best segmentation of the corpus using the estimated segment quality scores; and (4) compute rectified features using the segmented corpus and repeat steps (2)–(4) until the result converges. Figure 4.8 shows the high/low quality POS patterns learned using entity names found in the corpus.

	POS Tag Pattern	Example
Good (high score)	*NNP NNP* *NN NN* *CD NN* *JJ NN*	San Francisco/Barack Obama/United States comedy drama/car accident/club captain seven network/seven dwargs/2001 census crude oil/nucletic acid/baptist church
Bad (low score)	*DT JJ NND* *CD CD NN IN* *NN IN NNP NNP* *WD RB IN*	a few miles/the early stages/the late 1980s 2 : 0 victory over/1 : 0 win over rating on rotten tomatoes worked together on/spent much of

Figure 4.8: Example POS tag patterns.

After generating the set of candidate relation mentions from the detected candidate entity mentions, the authors propose to apply network embedding to help infer entity and relation types. Intuitively, two relation mentions sharing many text features (i.e., with similar distribution over the set of text features including head token, POS tags, entity mention order, etc.) likely have similar relation types; and text features co-occurring with many relation mentions in the corpus tend to represent close type semantics. For example, in Figure 4.9, ("Barack Obama," "US," "S1") and ("Barack Obama," "United States," "S3") share multiple features including context word 'president' and first entity mention argument "Barack Obama," and thus they are likely of the same relation type (i.e., "president_of").

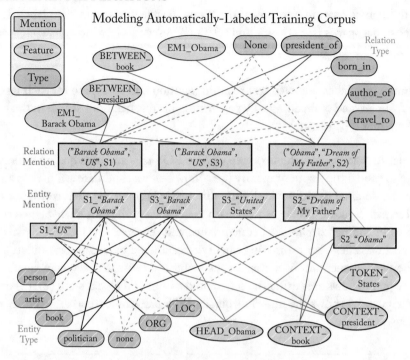

Figure 4.9: Network formed by entity/relation mentions, features, and types.

By further modeling the noisy distant labels from knowledge base and enforcing the additive operation[5] in the vector space, a joint optimization objective function is formulated to learn embeddings for both relation/entity mentions and relation/entity types.

It is reported in the paper that the effectiveness of joint extraction of typed entities and relations has been verified across different domains (e.g., news, biomedical), with an average of 25% improvement in F1 score compared to the next best method. Table 4.6 shows the output of the algorithm COTYPE together with other two competitive algorithms on two news sentences from the Wiki-KBP[6] dataset.

4.4   RESEARCH FRONTIER

There are many unexplored territories and challenging research issues. Here we outline some of the research directions stemming from our work. They share the common research goal that the mined phrases will become more task relevant and carry more semantics.

---

[5]In a relation mention $z = (m_1, m_2, s)$, embedding vector of m_1 should be a nearest neighbor of the embedding vector of m_2 plus the embedding vector of relation mention z.

[6]It uses 1.5M sentences sampled from 780,000 Wikipedia articles as training corpus and 14k manually annotated sentences from 2013 KBP slot filling assessment results as test data.

Table 4.6: Example output of COTYPE and the compared methods on two news sentences from the Wiki-KBP dataset. r^* stands for relation type and $\mathcal{Y}^*$ stands for entity type.

Text	*Blake Edwards*, a prolific filmmaker who kept alive the tradition of slapstick comedy, died Wednesday of pneumonia at a hospital in *Santa Monica*.	*Anderson* is survived by his wife Carol, sons *Lee* and Albert, daughter Shirley Englebrecht and nine grandchildren.
MultiR Hoffmann et al. [2011]	r^*: person:country_of_birth, $\mathcal{Y}^*_1$: {N/A}, $\mathcal{Y}^*_2$: {N/A}	r^*: None, $\mathcal{Y}^*_1$: {N/A}, $\mathcal{Y}^*_2$: {N/A}
Logistic Mintz et al. [2009]	r^*: per:country_of_birth, $\mathcal{Y}^*_1$: {person}, $\mathcal{Y}^*_2$: {country}	r^*: None, $\mathcal{Y}^*_1$: {person}, $\mathcal{Y}^*_2$: {person, politician}
CoType	r^*: person:place_of_death, $\mathcal{Y}^*_1$: {person,artist,director}, $\mathcal{Y}^*_2$: {location, city}	r^*: person:children, $\mathcal{Y}^*_1$: {person}, $\mathcal{Y}^*_2$: {person}

1. **Multi-sense Phrase Mining.** During the process of phrase mining, we typically assume phrase is represented as continuous word sequence. Our next challenge is to explore the underlying concept for each phrase mention and further rectify the phrase frequency. Such refinement encounters two problems: (1) variety: many phrase mentions may refer to the same concept (e.g., "page rank" and "PageRank," cytosol and cytoplasm); and (2) ambiguity: multiple concepts may share the same phrase mentions (e.g., PCR can refer to polymerase chain reaction or principle component regression). Such refinement is easier to achieve from the perspective of phrase evolving and contextual topic. When relational database was first introduced in 1970, "data base" was a simple composition of two words, and then with its gained popularity people even invented a new word "database," clearly as a whole semantic unit. In the context of machine learning, PCR certainly refers to principle component regression instead of polymerase chain reaction.

2. **Phrase Mining For Users.** This book mainly introduce the techniques for extracting phrases from documents. It can often be observed that unstructured textual data and users are interconnected, particularly in the big data era where social network and user-created content become popular. Together with mining massive unstructured data, one can expect to create profiles for users in the format of salient phrases by analyzing his/her activities, which can be utilized for future recommendation and behavior analysis. One promising solution is to model the user-content interaction as information network in the sense that links connect different types of nodes such as documents, keyphrases, and users. Such data model allows information propagation and many network-based algorithms can be applied.

3. **Phrase Mining For Fresh Content.** All the methods discussed in this book are data-driven and rely on frequent phrase mentions to certain extend. Accordingly, a large-scale dataset is necessary due to the data redundancy. On the other hand, the same philosophy is not suitable for fresh content when a new phrase is just formed. Instead of purely depending on the phrase mentions, contextual knowledge such as "Hearst patterns" is also useful. It is certainly an open problem to learn these textual patterns automatically and effectively. As time goes by, statistics of a phrase will eventually be sufficient, allowing our proposed methods to prove its power. It is interesting to see how much a hybrid method can benefit from this scenario as well.

Bibliography

Khurshid Ahmad, Lee Gillam, and Lena Tostevin. University of surrey participation in trec8: Weirdness indexing for logical document extrapolation and retrieval (wilder). In *TREC*, 1999. 21

Helena Ahonen. Knowledge discovery in documents by extracting frequent word sequences. *Library Trends*, 48(1), 1999. 5

Armen Allahverdyan and Aram Galstyan. Comparative analysis of viterbi training and maximum likelihood estimation for hmms. In *Advances in Neural Information Processing Systems 24*, pages 1674–1682, 2011. 15, 16

David Arthur and Sergei Vassilvitskii. k-means++: The advantages of careful seeding. In *Proc. of the 18th Annual ACM-SIAM Symposium on Discrete Algorithms*, pages 1027–1035, 2007.

Srikanta Bedathur, Klaus Berberich, Jens Dittrich, Nikos Mamoulis, and Gerhard Weikum. Interesting-phrase mining for ad-hoc text analytics. *Proc. of the VLDB Endowment*, 3(1-2): 1348–1357, 2010. DOI: 10.14778/1920841.1921007. 29

Christopher M Bishop. *Pattern Recognition and Machine Learning*. Springer-Verlag New York, Inc., 2006. 15, 45

Christian Bizer, Jens Lehmann, Georgi Kobilarov, Sören Auer, Christian Becker, Richard Cyganiak, and Sebastian Hellmann. Dbpedia-a crystallization point for the web of data. *Web Semantics: Science, Services and Agents on the World Wide Web*, 7(3):154–165, 2009. DOI: 10.1016/j.websem.2009.07.002. 56

David M Blei, Andrew Y Ng, and Michael I Jordan. Latent dirichlet allocation. *Journal of Machine Learning Research*, 3:993–1022, 2003. 55, 60

Leo Breiman. Randomizing outputs to increase prediction accuracy. *Machine Learning*, 40(3): 229–242, 2000. Springer. 39

Leo Breiman. Random forests. *Machine Learning*, 45(1):5–32, 2001. 13

Kuang-hua Chen and Hsin-Hsi Chen. Extracting noun phrases from large-scale texts: A hybrid approach and its automatic evaluation. In *Proc. of the 32nd Annual Meeting on Association for Computational Linguistics*, pages 234–241, 1994. DOI: 10.3115/981732.981764. 5

Gregory F Cooper. The computational complexity of probabilistic inference using Bayesian belief networks. *Artificial Intelligence*, 42(2):393–405, 1990. DOI: 10.1016/0004-3702(90)90060-d. 59

Mohamed Yehia Dahab, Hesham A Hassan, and Ahmed Rafea. Textontoex: Automatic ontology construction from natural english text. *Expert Systems with Applications*, 34(2):1474–1480, 2008. DOI: 10.1016/j.eswa.2007.01.043. 59

Marina Danilevsky, Chi Wang, Nihit Desai, Xiang Ren, Jingyi Guo, and Jiawei Han. Automatic construction and ranking of topical keyphrases on collections of short documents. In *Proc. of the SIAM International Conference on Data Mining*, 2014. DOI: 10.1137/1.9781611973440.46. 6

Marie-Catherine De Marneffe, Bill MacCartney, Christopher D Manning, et al. Generating typed dependency parses from phrase structure parses. In *Proc. of the 5th International Conference on Language Resources and Evaluation*, volume 6, pages 449–454, 2006. 47

Paul Deane. A nonparametric method for extraction of candidate phrasal terms. In *Proc. of the 43rd Annual Meeting on Association for Computational Linguistics*, pages 605–613, 2005. DOI: 10.3115/1219840.1219915. 35, 36

Scott C. Deerwester, Susan T. Dumais, Thomas K. Landauer, George W. Furnas, and Richard A. Harshman. Indexing by latent semantic analysis. *Journal of the American Society for Information Science*, 41(6):391–407, 1990. DOI: 10.1002/(sici)1097-4571(199009)41:6%3C391::aid-asi1%3E3.0.co;2-9. 55

Ahmed El-Kishky, Yanglei Song, Chi Wang, Clare R Voss, and Jiawei Han. Scalable topical phrase mining from text corpora. *Proc. of the VLDB Endowment*, 8(3), 2015. DOI: 10.14778/2735508.2735519. 10, 35, 36, 47, 60

Geoffrey Finch. *Linguistic Terms and Concepts*. Macmillan Press Limited, 2000. DOI: 10.1007/978-1-349-27748-3. 37, 43

Katerina Frantzi, Sophia Ananiadou, and Hideki Mima. Automatic recognition of multi-word terms: The c-value/nc-value method. *International Journal on Digital Libraries*, 3(2):115–130, 2000. DOI: 10.1007/s007999900023. 5, 10, 35

Evgeniy Gabrilovich and Shaul Markovitch. Computing semantic relatedness using wikipedia-based explicit semantic analysis. In *Proc. of the 20th International Joint Conference on Artificial Intelligence*, pages 1606–1611, 2007. 55

Chuancong Gao and Sebastian Michel. Top-k interesting phrase mining in ad-hoc collections using sequence pattern indexing. In *Proc. of the 15th International Conference on Extending Database Technology*, pages 264–275, 2012. DOI: 10.1145/2247596.2247628. 29

Pierre Geurts, Damien Ernst, and Louis Wehenkel. Extremely randomized trees. *Machine Learning*, 63(1):3–42, 2006. Springer. 39

Thomas Gottron, Maik Anderka, and Benno Stein. Insights into explicit semantic analysis. In *Proc. of the 20th ACM International Conference on Information and Knowledge Management*, pages 1961–1964, 2011. DOI: 10.1145/2063576.2063865. 55

Ziyu Guan, Long Chen, Wei Zhao, Yi Zheng, Shulong Tan, and Deng Cai. Weakly-supervised deep learning for customer review sentiment classification. In *Proc. of the 25th International Joint Conference on Artificial Intelligence*, pages 3719–3725, 2016. 60

Huan Gui, Jialu Liu, Fangbo Tao, Meng Jiang, Brandon Norick, and Jiawei Han. Large-scale embedding learning in heterogeneous event data. In *Proc. of the IEEE International Conference on Data Mining*, 2016. DOI: 10.1109/icdm.2016.0111. 67

John A Hartigan and Manchek A Wong. Algorithm as 136: A k-means clustering algorithm. *Journal of the Royal Statistical Society. Series C (Applied Statistics)*, 28(1):100–108, 1979. DOI: 10.2307/2346830.

Samer Hassan and Rada Mihalcea. Semantic relatedness using salient semantic analysis. In *Proc. of the 25th AAAI Conference on Artificial Intelligence*, pages 884–889, 2011. 55

Raphael Hoffmann, Congle Zhang, Xiao Ling, Luke Zettlemoyer, and Daniel S Weld. Knowledge-based weak supervision for information extraction of overlapping relations. In *Proc. of the 49th Annual Meeting of the Association for Computational Linguistics: Human Language Technologies*, volume 1, pages 541–550, 2011.

Terry Koo, Xavier Carreras Pérez, and Michael Collins. Simple semi-supervised dependency parsing. In *46th Annual Meeting of the Association for Computational Linguistics*, pages 595–603, 2008. 5

Roger Levy and Christopher Manning. Is it harder to parse Chinese, or the Chinese treebank? In *Proc. of the 41st Annual Meeting on Association for Computational Linguistics*, volume 1, pages 439–446, 2003. DOI: 10.3115/1075096.1075152. 47

Yanen Li, Bo-Jun Paul Hsu, ChengXiang Zhai, and Kuansan Wang. Unsupervised query segmentation using clickthrough for information retrieval. In *Proc. of the 34th International ACM SIGIR Conference on Research and Development in Information Retrieval*, pages 285–294, 2011. DOI: 10.1145/2009916.2009957. 15

Jialu Liu, Jingbo Shang, Chi Wang, Xiang Ren, and Jiawei Han. Mining quality phrases from massive text corpora. In *Proc. of the ACM SIGMOD International Conference on Management of Data*, pages 1729–1744, 2015. DOI: 10.1145/2723372.2751523. 36, 47, 55

Jialu Liu, Xiang Ren, Jingbo Shang, Taylor Cassidy, Clare R. Voss, and Jiawei Han. Representing documents via latent keyphrase inference. In *Proc. of the 25th International Conference on World Wide Web*, pages 1057–1067, 2016. DOI: 10.1145/2872427.2883088. 55, 56, 59

Gonzalo Martínez-Muñoz and Alberto Suárez. Switching class labels to generate classification ensembles. *Pattern Recognition*, 38(10):1483–1494, 2005. Elsevier. 39

Ryan McDonald, Fernando Pereira, Kiril Ribarov, and Jan Hajič. Non-projective dependency parsing using spanning tree algorithms. In *Proc. of the Conference on Human Language Technology and Empirical Methods in Natural Language Processing*, pages 523–530, 2005. DOI: 10.3115/1220575.1220641. 5

Rada Mihalcea and Paul Tarau. Textrank: Bringing order into texts. In *Proc. of the Conference on Empirical Methods in Natural Language Processing*, 2004. 47

Tomas Mikolov, Ilya Sutskever, Kai Chen, Greg S Corrado, and Jeff Dean. Distributed representations of words and phrases and their compositionality. In *Advances in Neural Information Processing Systems 26*, pages 3111–3119, 2013. 29, 60

Mike Mintz, Steven Bills, Rion Snow, and Dan Jurafsky. Distant supervision for relation extraction without labeled data. In *Proc. of the Joint Conference of the 47th Annual Meeting of the ACL and the 4th International Joint Conference on Natural Language Processing of the AFNLP*, volume 2, pages 1003–1011, 2009. DOI: 10.3115/1690219.1690287.

Joakim Nivre, Marie-Catherine de Marneffe, Filip Ginter, Yoav Goldberg, Jan Hajic, Christopher D Manning, Ryan McDonald, Slav Petrov, Sampo Pyysalo, Natalia Silveira, et al. Universal dependencies v1: A multilingual treebank collection. In *Proc. of the 10th International Conference on Language Resources and Evaluation*, pages 1659–1666, 2016. 47

Deepak Padmanabhan, Atreyee Dey, and Debapriyo Majumdar. Fast mining of interesting phrases from subsets of text corpora. In *Proc. of the 17th International Conference on Extending Database Technology*, pages 193–204, 2014. 29

Aditya Parameswaran, Hector Garcia-Molina, and Anand Rajaraman. Towards the web of concepts: Extracting concepts from large datasets. *Proc. of the VLDB Endowment*, 3(1-2):566–577, 2010. DOI: 10.14778/1920841.1920914. 6, 35, 36, 41, 47

Youngja Park, Roy J Byrd, and Branimir K Boguraev. Automatic glossary extraction: Beyond terminology identification. In *Proc. of the 19th International Conference on Computational Linguistics*, volume 1, pages 1–7, 2002. DOI: 10.3115/1072228.1072370. 5, 10, 35

Vasin Punyakanok and Dan Roth. The use of classifiers in sequential inference. In *Advances in Neural Information Processing Systems 13*, pages 995–1001, 2001. 5

Xiang Ren, Yuanhua Lv, Kuansan Wang, and Jiawei Han. Comparative document analysis for large text corpora. In *Proc. of the 10th ACM International Conference on Web Search and Data Mining*, 2017a. DOI: 10.1145/3018661.3018690. 63

Xiang Ren, Zeqiu Wu, Wenqi He, Meng Qu, Clare Voss, Heng Ji, Tarek Abdelzaher, and Jiawe Han. Cotype: Joint extraction of typed entities and relations with knowledge bases. In *Proc. of the 26th International Conference on World Wide Web*, 2017b. 68

Mark Sanderson and Bruce Croft. Deriving concept hierarchies from text. In *Proc. of the 22nd Annual International ACM SIGIR Conference on Research and Development in Information Retrieval*, pages 206–213, 1999. DOI: 10.1145/312624.312679. 59

Helmut Schmid. Probabilistic part-of-speech tagging using decision trees. In *New Methods in Language Processing*, page 154, 2013. 36

Jingbo Shang, Jialu Liu, Meng Jiang, Xiang Ren, Clare Voss, and Jiawei Han. Automated Phrase Mining from Massive Text Corpora, *arXiv* preprint arXiv:1702.04457v1, 2017. 55

Alkis Simitsis, Akanksha Baid, Yannis Sismanis, and Berthold Reinwald. Multidimensional content exploration. *Proc. of the VLDB Endowment*, 1(1):660–671, 2008. DOI: 10.14778/1453856.1453929. 5

Yangqiu Song, Haixun Wang, Zhongyuan Wang, Hongsong Li, and Weizhu Chen. Short text conceptualization using a probabilistic knowledge base. In *Proc. of the 22nd International Joint Conference on Artificial Intelligence*, pages 2330–2336, 2011. DOI: 10.5591/978-1-57735-516-8/IJCAI11-388. 55

Fangbo Tao, Honglei Zhuang, Chi Wang Yu, Qi Wang, Taylor Cassidy, Lance Kaplan, Clare Voss, and Jiawei Han. Multi-dimensional, phrase-based summarization in text cubes. *Data Engineering*, 39(3):74–84, 2016. 63

Beidou Wang, Can Wang, Jiajun Bu, Chun Chen, Wei Vivian Zhang, Deng Cai, and Xiaofei He. Whom to mention: Expand the diffusion of tweets by @ recommendation on micro-blogging systems. In *Proc. of the 22nd International Conference on World Wide Web*, pages 1331–1340, 2013. DOI: 10.1145/2488388.2488505. 60

Chi Wang, Wei Chen, and Yajun Wang. Scalable influence maximization for independent cascade model in large-scale social networks. *Data Mining and Knowledge Discovery*, 25(3):545–576, 2012. DOI: 10.1007/s10618-012-0262-1. 59

Ian H Witten, Gordon W Paynter, Eibe Frank, Carl Gutwin, and Craig G Nevill-Manning. Kea: Practical automatic keyphrase extraction. In *Proc. of the 4th ACM Conference on Digital Libraries*, pages 254–255, 1999. DOI: 10.1145/313238.313437. 47, 58

Wentao Wu, Hongsong Li, Haixun Wang, and Kenny Q Zhu. Probase: A probabilistic taxonomy for text understanding. In *Proc. of the ACM SIGMOD International Conference on Management of Data*, pages 481–492, 2012. DOI: 10.1145/2213836.2213891. 56

Endong Xun, Changning Huang, and Ming Zhou. A unified statistical model for the identification of English basenp. In *Proc. of the 38th Annual Meeting on Association for Computational Linguistics*, pages 109–116, 2000. DOI: 10.3115/1075218.1075233. 5

Xiaoxin Yin and Sarthak Shah. Building taxonomy of web search intents for name entity queries. In *Proc. of the 19th International Conference on World Wide Web*, pages 1001–1010, 2010. DOI: 10.1145/1772690.1772792. 59

Ziqi Zhang, José Iria, Christopher A Brewster, and Fabio Ciravegna. A comparative evaluation of term recognition algorithms. *Proc. of the 6th International Conference on Language Resources and Evaluation*, 2008. 5, 35

Authors' Biographies

JIALU LIU

Jialu Liu, an engineer at Google Research in New York, is working on structured data for knowledge exploration. He received his B.Sc. from Zhejiang University, China, in 2007 and Ph.D. degree in computer science from the University of Illinois at Urbana-Champaign in 2015. His research has been focused on scalable data mining, text mining, and information extraction.

JINGBO SHANG

Jingbo Shang, is a Ph.D. candidate in the Department of Computer Science at the University of Illinois at Urbana-Champaign. He received a B.Sc. from Shanghai Jiao Tong University, China in 2014. His research focuses on mining and constructing structured knowledge from massive text corpora.

JIAWEI HAN

Jiawei Han, Abel Bliss Professor, Department of Computer Science, the University of Illinois, has been researching data mining, information network analysis, and database systems, and has been involved in over 700 publications. He served as the founding Editor-in-Chief of *ACM Transactions on Knowledge Discovery from Data (TKDD)*. Jiawei received the ACM SIGKDD Innovation Award (2004), IEEE Computer Society Technical Achievement Award (2005), and IEEE Computer Society W. Wallace McDowell Award (2009). He is a Fellow of ACM and a Fellow of IEEE. His co-authored textbook, *Data Mining: Concepts and Techniques* (Morgan Kaufmann), has been adopted worldwide.